# Restoration ecology

# Restoration ecology

## a synthetic approach to ecological research

edited by

*William R. Jordan III*
University of Wisconsin-Madison Arboretum, Wisconsin, USA

*Michael E. Gilpin*
Biology Department, University of California-San Diego, California, USA

*John D. Aber*
Forestry Department, University of Wisconsin-Madison, Wisconsin, USA

CAMBRIDGE
UNIVERSITY PRESS

Published by the Press Syndicate of the University of Cambridge
The Pitt Building, Trumpington Street, Cambridge CB2 1RP
40 West 20th Street, New York, NY 10011–4211, USA
10 Stamford Road, Oakleigh, Victoria 3166, Australia

First published 1987
Reprinted 1989
First paperback edition 1990
Reprinted 1992

Printed in Great Britain at the University Press, Cambridge

*British Library cataloguing in publication data*

Restoration ecology: a synthetic approach
to ecological research.
1. Ecology
I. Jordan, William R.    II. Gilpin, Michael E.
III. Aber, John D.
574.5    QH541

*Library of Congress cataloguing in publication data*

Restoration ecology.
1. Ecology. 2. Reclamation of land. 3. Nature
conservation. I. Jordan, William R. II. Gilpin,
Michael E. III. Aber, John D.
QH541.R456    1987    574.5    86-33401

ISBN 0 521 33110 2 hardback
ISBN 0 521 33728 3 paperback

# Contents

# Contributors

J. D. Aber
Dept. of Forestry
University of Wisconsin–Madison
Madison, WI 53706, USA

W. H. Adey
Marine Systems Laboratories
MNH 310
Smithsonian Institution
Washington DC 20008, USA

T. F. H. Allen
Dept. of Botany
University of Wisconsin–Madison
Madison, WI 53706, USA

W. C. Ashby
Dept. of Botany
Southern Illinois University
Carbondale, IL 62901, USA

F. H. Bormann
School of Forestry &
Environmental Studies
Yale University
New Haven, CT 06511, USA

A. D. Bradshaw
Dept. of Botany
University of Liverpool
Liverpool, L69 3BX, UK

J. Cairns, Jr
University Center for
Environmental Studies
Virginia Polytechnic Institute
Blacksburg, VA 24061, USA

G. D. Cooke
Dept. of Biological Sciences
Kent State University
Kent, OH 44242, USA

G. Cottam
Dept. of Botany
University of Wisconsin–Madison
Madison, WI 53706, USA

J. Diamond
Dept. of Physiology
University of California
Los Angeles, CA 90024, USA

J. J. Ewel
Dept. of Botany
Bartram Hall
University of Florida
Gainesville, FL 32611, USA

M. E. Gilpin
Dept. of Biology
University of California
La Jolla, CA 92093, USA

K. L. Gross
Dept. of Botany
Ohio State University
Columbus, OH 43210, USA

J. Harper
School of Plant Biology
University College of North Wales
Bangor, Gwynedd LL57 2UW, UK

T. W. Hoekstra
Rocky Mountain Forest & Range
Experimental Station
Ft. Collins, CO 80521, USA

E. A. Howell
Dept. of Landscape Architecture
University of Wisconsin–Madison
Madison, WI 53706, USA

W. R. Jordan III
UW-Madison Arboretum
1207 Seminole Highway
Madison, WI 53711, USA

J. A. MacMahon
Dept. of Biology
Utah State University
Logan, UT 84321, USA

T. McNeilly
Dept. of Botany
University of Liverpool
Liverpool, L69 3BX, UK

R. M. Miller
Land Reclamation Program
Argonne National Laboratory
Argonne, IL 60439, USA

E. J. Morton
Dept. of Zoological Research
National Zoological Park
Smithsonian Institution
Washington, DC 20008, USA

J. Powers
Prairie Ridge Nursery
9738 Overland Road
Mt Horeb, WI 53572, USA

V. M. Kline
UW-Madison Arboretum
1207 Seminole Highway
Madison, WI 53711, USA

M. L. Rosenzweig
Dept. of Ecology
University of Arizona
Tucson, AZ 85721, USA

F. Turner
School of Arts & Humanities
University of Texas–Dallas
Richmond, TX 75080, USA

E. B. Welch
Dept. of Civil Engineering
University of Washington
Seattle, WA 98195, USA

P. Werner
Tropical Ecosystems Res. Centre,
CSIRO
Darwin Laboratories
PM B44
Winnellie, Northern Terr.
Australia 5789

# Acknowledgements

This book represents the generous efforts of a number of people whose names are not represented in the table of contents.

To begin with, the symposium on which the book is based would not have taken place without the leadership and support of the University of Wisconsin–Madison Arboretum Director, Gregory Armstrong, who initiated planning for this program within weeks after assuming his post in 1983, and who has enthusiastically supported every aspect of the project, from planning the meeting to the editing of the resulting manuscripts.

Special thanks are also due to Chairman Gerald Gerloff, Donald Waller, and other members of the Research Subcommittee, which planned the meeting, the Arboretum Committee, which sponsored it, and to former Arboretum Naturalist James Zimmerman and Arboretum Rangers Keith Wendt and Brock Woods for valuable discussions of some of the ideas developed in this book.

We are grateful, too, to Arboretum Secretary Nancy Dopkins for typing, retyping and checking manuscripts, to Barbara Jordan for checking the final drafts and to Pam Holton for her skilful rendition of tables and graphs.

We also thank Peg Watrous for help in checking chapter references and members of the Friends of the Arboretum for assistance in planning and carrying out the symposium, and for help in proofreading.

Funding for the symposium was provided by the Knapp Fund, the University of Wisconsin Graduate School, a UW Anonymous Fund and the UW Arboretum.

# I | Introduction

In the planting of the seeds of most trees, the best gardeners do no more than follow nature, though they may not know it...So, when we experiment in planting forests, we find ourselves at last doing as Nature does. Would it not be well to consult with Nature in the outset? for she is the most extensive and experienced planter of us all...

Henry David Thoreau, *On the Succession of Forest Trees*

*William R. Jordan III*
University of Wisconsin-Madison Arboretum

*Michael E. Gilpin*
Department of Biology, University of California-San Diego

*John D. Aber*
Department of Forestry, University of Wisconsin-Madison

# 1 | Restoration ecology: ecological restoration as a technique for basic research

Just over fifty years ago, in the fall of 1935, a small contingent of Civilian Conservation Corps workers began replanting tallgrass prairie on a played-out piece of farmland on the edge of Madison that had recently been acquired by the University of Wisconsin for an arboretum. The project, which was to result in the creation of the 24 ha prairie that is today the esthetic and ecological centerpiece of the UW Arboretum, was carried out under the direction of Aldo Leopold, and that same season, against a background of catastrophic drought and economic depression, Leopold himself undertook the restoration of another piece of derelict farmland near the sandy banks of the Wisconsin River, just an hour's drive north of the city (Fig. 1.1).

These two projects – one carried out in the context of a university and a federal work relief program, the other carried out by a single man musing over his shovel on weekends – marked a beginning and a signal moment in the conservation movement of which Leopold was both a prophet and a pioneer. For the idea of ecological restoration was a new one in 1935. Elsewhere, across the heartland of the continent, CCC boys were at work helping to repair the ravages of an exploitative system of agriculture. But in the vast majority of cases these efforts involved mere revegetation of the landscape in an attempt to stabilize eroding soil, or ultimately to produce something of economic value, such as rangeland or a salable crop of timber. What was distinctive about the projects being carried out at the new arboretum and at Leopold's weekend retreat was that they were committed to the actual, painstaking *restoration* of the plant and animal communities native to the area. This was revegetation carried out in the most precise and meticulous possible imitation of nature, and so represented in its clearest form the novel combination of agriculture and ecology out of which has grown the science and art of ecological restoration.

This book is a direct outgrowth of the UW Arboretum's restoration efforts. It represents an attempt to define more clearly their meaning and significance and also to consider how the Arboretum might continue to contribute to the development of restoration, both as a form of environmental technology and as a technique for basic research.

At least some of the implications of the Arboretum experiment are now fairly obvious. In part a response to the ecological devastation of the Dust Bowl and the deforestation of the upper Great Lakes area, which was virtually complete by the 1930s, the Arboretum's restoration efforts have always stood as a dramatic testimony to the possibility of a positive, healthy relationship between human beings and the land. In other words, the Arboretum has turned out to be a proving ground for the development of a highly practical expression of Leopold's own land ethic. Inadequate and unfinished as the Arboretum's restored communities may be in their various ecological deficiencies, their missing species, their variety of unwanted exotics, their recently disturbed soils, their youth and the simple fact that they are all quite small, they do represent a kind of environmental triumph – a clear sign that it is possible, both technically and socially, to reverse environmental destruction, to go from pasture or eroded and played-out plowland back to prairie or to forests of maple, spruce and pine.

At the same time, the development of ecological restoration, in which the rather unimposing efforts at the UW Arboretum played a pioneering role, raises certain questions, and these have become increasingly urgent as the practice of restoration has become more and more widespread. What, for example, is the value of restoration as a way of dealing with environmental problems? What role might the deliberate restoration of disturbed ecosystems be expected to play in the conservation of habitat, of rare and endangered species and of biological diversity generally? To what extent will it be possible to restore authentic communities with complete species lists? Under what conditions will it be feasible to do so? Will society be willing to pay the cost of such ventures? More generally, what sort of relationship with nature does restoration signify and encourage? What might be its value as a way of developing a closer relationship between an industrial society and nature? In particular, what is its relationship with the activity that represents the central feature of our society's intellectual relationship with nature – that is, ecology, the science of nature seen as a whole?

All these questions are both interesting and important. It is only the last of them, however – the question of the relationship between the practice of restoration and the science of ecology – that we intend to explore in some detail in this book.

We are approaching this subject in a somewhat novel way. Usually it is assumed that restoration is a practical matter, a form of applied ecology, and it is taken for granted that it is the insights achieved as a result of more "basic" research that will provide the basis for successful restoration practice. Ecologists, in other words, are expected to do most of the talking, restorationists most of the listening.

*Fig. 1.1.* Late in the fall of 1935, Civilian Conservation Corps workers began turning over sod on an old pasture at the University of Wisconsin–Madison Arboretum in the early stages of restoring tallgrass prairie. Initiated for the purpose of creating representative ecological communities to serve as subjects for research, efforts like these eventually led to an awareness of the value of restoration itself as a way of approaching ecological research (UW Arboretum files).

In the discussion carried on in this book, we are making a deliberate attempt to turn this around. For one thing, we have systematically ignored the traditional distinction between theory and practice. But more than that we have attempted to *unite* the two traditions by drawing attention to the tremendous value of restoration, not only as a form of environmental technology, but also as a technique for basic research. The idea here is simply that one of the most valuable and powerful ways of studying something is to attempt to reassemble it, to repair it, and to adjust it so that it works properly. This idea is a commonplace in other areas of endeavor. In the chapters that follow we explore its application to ecological systems. Indeed, it is this idea of restoration (and ecological management generally) as a technique for raising basic questions and testing fundamental ideas, leading in turn to improved restoration and management techniques that we have termed restoration – or synthetic – ecology (Aber & Jordan, 1985).

This is the idea that runs through and unites all the subsequent chapters. Disparate as they may seem, coming as they do from a dozen different directions, dealing with a variety of systems from different points of view, and with different degrees of emphasis on "theory" and "practice", each of them has one thing in common with all the rest: each one deals with the reassembly, assembly or partial assembly of an ecological system and with what has been or might be learned from this process.

The point is that this is a book about a technique – not only about techniques for restoring systems, but *simultaneously* about a technique for doing ecological research. We have therefore tried to explore the value of this technique in dealing with as many areas and ecological issues as possible. Certainly the result is not a comprehensive survey of issues that might be approached in this way, but we hope it is at least representative. If we have not included every mountaintop that must be scaled, at least perhaps we have included the major mountain ranges.

The book itself is the outcome of a symposium held in Madison on October 11 and 12, 1984, as one of several events marking the Arboretum's fiftieth anniversary. This meeting was attended by more than 300 persons from all over the US and Canada. It was well received, and it is gratifying to report that it has turned out to be an event of some importance for the Arboretum. In addition to the stimulating influence of the meeting itself and its many enthusiastic participants, it has led directly to the establishment of a Center for Restoration Ecology, a coalition of faculty and staff headquartered at the Arboretum, and also to a plan to create an endowment to support research in restoration ecology at the Arboretum. This endowment will eventually support a professorship – the Aldo Leopold Professor of Restoration

Ecology – and several research assistantships, all committed to studies in this rapidly developing discipline.

Just what, then, is the relationship between the practice of ecological restoration and the science of ecology? What has it been? What might it be? More than that, what *ought* it to be?

At least one answer to these questions is fairly obvious. It was spelled out by Leopold himself in a short speech at the dedication of the Arboretum in 1934. In this speech, which seems to have been the first public description of the Arboretum's novel restoration plan, Leopold indicated that the restoration was definitely being carried out for ecological purposes. Specifically, it was being done so that the various communities that would be restored, many of which had become rare since the time of settlement, would be available for ecologists to study.

"The time has come", Leopold argued, "for science to busy itself with the earth itself. The first step is to reconstruct a sample of what we had to begin with" (Leopold, 1934).

At this point, Leopold placed little emphasis on the proposed restoration effort itself, referring to it as merely a "first step" and "a starting point in the long and laborious job of building a permanent and mutually beneficial relationship between civilized men and a civilized landscape".

As the project got under way, however, it soon became clear that restoration was somewhat more than a "first step" towards a better relationship with the environment and a deeper understanding of it. It was itself a significant challenge, one which was not merely a prelude to ecological research, but which went hand in hand with it.

Perhaps the clearest example of this relationship between ecological theory and restoration practice to come out of the early work at the Arboretum was the research on the ecological role of fire in prairies (Fig. 1.2). When prairie restoration had begun at the Arboretum in the 1930s, it was known that fires had been frequent on the prairies of the Midwest in presettlement times. Little was known, however, about the ecological significance of fire in grassland communities. It had been the failure of the Arboretum's early attempts to restore prairie *without* fire that had led to a classic series of experiments on the role of fire in prairie ecology which was carried out by John Curtis and Max Partch during the 1940s.

These experiments clearly established the importance of fire as a factor in the ecology of prairies and provided indispensable information for prairie restorationists and managers. They were also a significant contribution to ecologists' growing understanding of the importance of fire in plant communities. Even more generally, however, they pointed towards the value of restoration as a way of testing ideas about the communities being restored, and experiences of this kind did lead to a growing appreciation of the difficulty of restoration itself, and of its value at least as an adjunct to ecological research. "It is believed", Curtis and Partch wrote in their report of the early burning experiments, "that by the judicious use of fire or other biotic controls, fairly complete associations of each of the formations can be established and maintained within the 1400 acre limits of the Arboretum; *it is hoped that much information of value concerning the dynamics of formation boundaries can be obtained in the course of such establishment*" (emphasis added) (Curtis & Partch, 1948).

A decade later, commenting more generally on his experiences with the restoration of ecological communities at the Arboretum, Curtis wrote, "The problems encountered in the attempts at artificial establishment of natural plant communities have demonstrated the complexities of community

*Fig. 1.2.* Dismal results of the Arboretum's early attempt to restore prairie without using fire led to classic experiments on the role of fire in prairie ecology. Shown here are Aldo Leopold (second from left) and several students at an experimental burn on the Arboretum prairie in the 1940s (photograph courtesy of Robert McCabe).

integration in a way that no other experience could provide" (Curtis, 1959).

In fact the experience with fire on the prairies was only one example of a series of insights that had emerged from the Arboretum's restoration efforts. By the time the fifty years Leopold had envisioned for the restoration effort were up, it was possible to list several more such instances. Among them:

☐ The siting of communities on the basis of what was known at the time restoration began had turned out to provide, in effect, a series of tests of that knowledge, and as Anthony Bradshaw points out in Chapter 2, from this point of view, failures were at least as valuable as successes. Attempts to establish pine forests 100 km or more south of their native range, for example, had been only marginally successful. The pines grew well enough, but many understory species had not. Their poor performance raised questions about the factors influencing the development of understory in natural pine forests, and led to research which showed that a key factor in the poor growth of the understory plants at the Arboretum was a subtle difference between patterns of precipitation in northern and southern Wisconsin (Anderson, Loucks & Swain, 1969).

☐ Similarly, the placing of plants within communities had led to numerous insights into the dynamics and functioning of the communities. Some of these are described in subsequent chapters by Grant Cottam and John Aber.

☐ Reestablishment of communities on sites from which they had been removed many years earlier had provided opportunities for studies of recovery of the soil beneath them. Plots set up for this purpose by Curtis and Soils Scientist Francis Hole apparently represent the oldest plots of this kind anywhere, and 30 years later are still producing information about soil development that could not be obtained in any other way (Van Rooyen, 1973).

In other words, looking back over the first fifty years of restoration efforts at the Arboretum, it was possible to discern a pattern in the ecological research that had been carried out on the property, and in the relationship between this research and the restoration effort. While the original idea had been to restore the communities *in order to have them to study*, the process of restoration itself apparently being more or less taken for granted, in fact it had been the restoration, the bringing together of the communities and the ongoing tinkering with them, that had provided some of the most interesting opportunities for study. This was an observation of considerable importance to the Arboretum, since even after Leopold's "50 years" none of the restored communities in the collection was a finished replica of a natural model, acceptable for study on the same terms. Nevertheless, interesting things had been and were being learned from them – and many of these insights were a result, directly or indirectly, of the restoration effort itself.

At the very least this experience illustrated the value of restoration as an opportunity or occasion for basic ecological research. It seemed to us, however, that it was possible to take this idea a step farther. If ecologists had been able to learn a good deal about ecological communities as a result of efforts to restore them for purely practical reasons (in this case simply to have them to serve as subjects for research), it seemed reasonable to suppose that this approach might be even more effective if the restoration efforts were conceived from the outset specifically as experiments – not empirical experiments designed to find out what works under a given set of conditions, but experiments designed to test basic ideas about why certain techniques work and how the community itself works.

This book is an attempt to explore the subject in some detail. It is best regarded, therefore, as a discussion from various points of view of several closely related questions about how ecology works and its relationship to the practice of ecological restoration. Just what, for example, has been learned as a result of attempts to restore or create ecological communities and ecosystems? What has been the exact nature of the relationship between restoration practice and ecological discovery? Has it been a necessary one, or merely fortuitous? To what extent is it possible to discern a *pattern* involving the use of restoration and synthesis of ecological systems, or of restored or artificial systems themselves, to raise questions and test fundamental ecological hypotheses? To what extent has this been a conscious process? And, finally, what might be the value of identifying this pattern and developing it more consciously and systematically as a way of carrying out and evaluating ecological research?

There are, of course, numerous precedents and even venerable authority for this approach to research, both in ecology and in science generally. Certainly, in the most general sense, there is nothing new about the idea of testing ideas about things by attempting to reassemble or repair them. This is something that craftsmen, inventors, tinkerers – and scientists – do all the time. Both Francis Bacon and René Descartes, the early theorists of scientific method, stressed not only the methods of analysis and distinction but also the complementary use of comparisons and a constructive – or synthetic – approach to the study of the complex.

A century later, this idea was expressed with particular clarity by Giovanni Battista Vico, an eighteenth century Italian philosopher and critic of Descartes, who, in his search for a way of distinguishing what can be known from what cannot, put forward "the doctrine that *verum et factum convertuntur*: that is, the condition of being able to know anything truly, to understand it as opposed to merely perceiving it, is that the knower himself should have made it" (Collingwood, 1946).

In fact, Vico's principle merely points to a familiar experience in science, technology; and, as Frederick Turner points out in Chapter 4, even in art: tinkering, trial and error and the imitation of nature, though most often associated with technology and engineering, are integral to the process of discovery and verification. Here, it is the metaphor of the machine that comes most readily to mind. We know that we understand a machine – a radio, for example, or a violin – when we can not only disassemble it, but reassemble it and adjust it so that it works properly. The idea is at least implicit in Leopold's own frequent use of this metaphor in his references to ecosystems and their management. And, not surprisingly, both Anthony Bradshaw and John Harper use the same metaphor in their discussions of the heuristic value of restoration, later in this section. The metaphor comes to mind, and clicks into place, because it is precisely the successful reconstruction, adjustment and repair of a device like a violin that is so convincing a demonstration of the tinkerer's skill.

This, of course, does not apply only to craftsmen and engineers. Essentially the same procedure is commonplace in the natural sciences as well. A chemist, for example, establishes the elemental composition of a substance by analysis, but demands synthesis as proof of structure. Similarly, biochemists and molecular biologists demonstrate their understanding of biological subsystems such as membranes and cellular organelles, and also processes such as photosynthesis, respiration and protein and DNA synthesis, by taking them or the subsystems responsible for them out of cells and then putting them back together and making them run *in vitro*. Implicit here is Vico's idea that one's understanding of a thing coincides with one's ability to create and recreate it – that we can claim to understand thoroughly only as far as we are able to reassemble, adjust and control. The principle is also suggested by work in organismal biology – especially with plants – in which scientists have learned to regenerate organisms from increasingly small subunits – first organs, then tissues, then cells and even naked protoplasts. In each case the new technique both depends on a deeper understanding of developmental processes and provides a way of testing and refining that understanding. It may be that the ultimate goal is to bring the whole structure together from its raw elements (Wilson, 1978). Wilson also points out that the experience of medicine supports the idea of the heuristic value of working with diseased and injured living systems in the attempt to heal them.

All these examples pertain only to non-living systems, or to biological systems at or below the organismal level. The question for us here is the extent to which this idea of the heuristic value of synthesis applies to ecology, which deals with nature at the higher levels of organization, the community and the ecosystem. Is the synthetic approach useful here, where interactions may be as complex as they are at the level of cell or organism, but where they may also be far less tightly integrated and are more subject to the influence of stochastic events?

Though it is impossible to be sure, we suspect the synthetic approach will prove at least as useful in ecology as in other areas. One reason analysis has been so useful in the physical sciences is that it is often possible to bring its subjects into the laboratory, to excise real and functioning subsystems, and to lay them out on the bench for detailed investigation. These subsystems cleave along recognized interfaces and, if carefully chosen, can often be regarded as typical. Because of this, the reassembly of such subsystems is often fairly straightforward, for their coupling occurs at limited and often easily defined boundaries. Hence their synthesis may not depend on new rules of science. It is taught as an art and is generally practiced for commercial gain rather than as an explicitly heuristic process.

On the ecological side of biology, on the other hand, at the level of population, community and ecosystem, the situation is perhaps different. Here, since the objects of study cannot be brought into the laboratory, there has always been a strong dependence on descriptive studies. But the amount the ecologist can learn simply by watching the system in its steady state is severely limited. Rather, as in the case of physical systems, there must be a response to external disturbances to expose the dynamic connections of the system. Moreover, because the responses of the system may be strongly non-linear, the perturbations must be extreme. As Mike Rosenzweig points out in Chapter 13, for example, populations have to be studied over the full range of densities, from zero to the equilibrium value. Similarly, individual species and assemblages must be studied not only at or near their preferred sites, where they tend to occur naturally, but on highly inappropriate sites as well (see, for example, Ch. 16). Since only the act of restoration can accomplish this, it may be that in ecology in particular, synthesis will someday become a coequal not only of description, but of analysis as well. And for some areas, such as studies of community structure, it could turn out to be the only feasible approach (Ch. 10).

In fact, the synthetic approach has been used regularly by ecologists from a very early period. The idea is latent, for example, in Thoreau's admonition, quoted at the beginning of this section, to consult nature in planting forests. Of course Thoreau was not an ecologist. But the essay on the succession of forest trees in which this remark occurs has long been considered a seminal ecological document. Similarly, some of Charles Darwin's ecological experiments and observations have about them a flavor of restoration ecology – his clearing of ground in order to study competition in the resulting community of weeds, for example, and his observations on the effects of planting trees on the development of a heath community (Darwin, 1859). In addition, of course, there is a long tradition within ecology of testing ideas synthetically, by assembling communities or constructing ecosystems – often highly simplified ones – under controlled conditions in the laboratory. Techniques of this kind have been used, for example, to study phenomena such as population dynamics, competition, predator-prey relationships and nutrient cycles. Clearly this is synthetic work, but it has rarely, if ever, been consciously linked with the synthetic work carried on in the field in the form of ecological restoration. In these terms, then, the question becomes: what might be the value of deliberately extending the carefully designed, synthetic experiment of the laboratory into the field, taking advantage of the heuristic value of the synthetic study in such a way as to gain real-world relevance, while retaining as much as possible the experimental rigor and precision of the laboratory experiment? (For an interesting discussion of the value of artificial "mesocosms" in ecological research, see Odum, 1984.)

All this suggests quite clearly the value of what might be called the synthetic approach to ecological research, of which restoration in its more familiar forms may be regarded simply as a special example. Fortunately, while this idea has apparently not been developed in detail, it has recently been discussed by several researchers representing a considerable range of expertise and professional experience. In a recent essay entitled *Wildlife management as ecological experimentation*, for example, four wildlife managers writing under a single pen name argue that, given the current state of their art, ecological management is for the most part not so much a routine procedure as a kind of ongoing experiment, or trial of ideas, and that it might best be practised in this spirit (McNab, 1983).

Management, of course, is a broader term than restoration, yet

"McNab's" comments are clearly relevant to restoration in the narrower sense as well. In any case, as we will discuss below, we are not concerned with *restoration* in the strictest sense, but more generally in the *control* of communities and ecosystems, of which restoration is only a particular example.

Equally to the point are two lines of reasoning recently put forward by two British ecologists, John Harper and Anthony Bradshaw. In an essay titled *After Description*, Harper, perhaps reflecting on his own early experience in agricultural research, argued, in effect, that agriculture provides a kind of model for the development of ecology as an experimental science (Harper, 1982). He pointed out that agriculture, as a result of its efforts to devise artificial ecosystems and to manage them with confidence, has already provided ecology with some of its most fundamental ideas. He then went on to argue for the value of a similar kind of piece-by-piece disassembly and reassembly – in other words synthesis or, ultimately, restoration – of ecological systems as a way of studying them. Here, it would seem, agriculture and ecology join in the form of restoration conceived both as a practical venture (like the more traditional forms of agriculture) and as a technique for basic research. Agriculture, in other words, is proposed as a paradigm for ecological research, the ecologist testing ideas about the more complex natural systems by attempting to create and manage them in the same way farmers manage their crops. This is precisely what we mean by the term restoration ecology.

In a similar vein, Bradshaw, one of the small handful of academic ecologists who have taken a serious interest in restoration as an intellectual challenge, recently suggested that, far from being simply a form of "applied" ecology, restoration actually represents an "acid test" of ecological ideas. "The acid test of our understanding", he writes, "is not whether we can take ecosystems to bits on pieces of paper, however scientifically, but whether we can put them together in practice and make them work. We can examine our competence by looking at the various problems in turn" (Bradshaw, 1983).

While both Harper's statement and Bradshaw's are essentially versions of the idea we propose to explore in this book, they are basically complementary in emphasis, Harper stressing the value of restoration or reassembly as a technique for research, Bradshaw stressing the ability to restore successfully as the ultimate test of understanding. Both Harper and Bradshaw have contributed to this volume further thoughts on these ideas, and to a considerable extent it will be the purpose of this book to consider them from various points of view and to explore their implications both for ecology and for the practical business of ecological restoration and management.

We think this is worth doing for several reasons:

☐ If restoration really is a valuable technique for ecological research, it will certainly be worthwhile to recognize this, to explore its strengths and weaknesses, and perhaps to exploit it more systematically as a research technique.

☐ This is especially true to the extent that Bradshaw's characterization of restoration as the crucial test of ecological understanding is valid. An immediate corollary of this idea, it seems to us, is that, far from being of peripheral interest to ecology, restoration actually deserves to be regarded as an organizing principle for ecological research, the basis for deciding which questions are most worth answering and which ones are irrelevant or of marginal interest. In other words, it suggests that restoration might provide the basis for organizing, evaluating, and even criticizing, ecological research. (An excellent example of the value of restoration as a basis for organizing ecological research is suggested by the early work at the UW Arboretum, where the need for detailed information about the native communities which were being restored was at least one motive for John Curtis's comprehensive studies of the vegetation of the state described in *The Vegetation of Wisconsin*; Curtis, 1959.)

☐ A second corollary of this idea is that, at least in ecology, the objectives of the theorist and the practitioner, though perhaps not identical, are certainly convergent. Both the restorationist and the restoration ecologist seek to reconstruct the system – the one in order to conserve it, the other in order to test ideas or to demonstrate an understanding of it. Recognizing this, and taking advantage of it, might provide a solid basis for a closer, two-way relationship between theory and practice in this area – a relationship similar to the one that exists in medicine, where clinical work and basic research often proceed hand in hand.

☐ Finally, by accepting restoration as its central challenge, ecology acquires not only an identifiable goal (understanding the system and being able to demonstrate this understanding in an objective, unambiguous way), but also a mission (being able to heal the system). Since the two are more or less congruent, this links ecology decisively to the land ethic. It also represents ecology truly functioning and thinking of itself as one of the healing arts.

In short, to the extent this line of reasoning is both sound and fruitful, restoration, properly understood, is an activity that belongs not on the margins of ecology, but at its center.

Of course, none of this makes much sense if one regards restoration merely as a form of "glorified gardening", and clearly it is necessary here to be very clear about what we mean by the term "restoration". For one thing, as we have already noted, the heuristic value of synthesis is most commonly discussed in terms of inanimate, physical objects, most frequently mechanical devices. The feeling is that if we can reassemble such a device and adjust it properly, then perhaps we can claim to understand it. The idea works, however, at least partly because the system referred to is inanimate and therefore has no power of self-repair. Living systems, in contrast, do have considerable powers of self-repair, and this must to some extent reduce the value of restoration as a crucial test of understanding. (Bradshaw discusses this point in some detail in Ch. 5.) Indeed, one reason ecologists have not taken restoration more seriously as a test of ecological ideas is no doubt because the goals of restoration projects are usually fairly broadly defined, and also because restorationists typically rely heavily on natural forces in achieving them. Restoration efforts, for example, often involve bringing in certain key "ingredients", then letting nature take its course in shaping the result. At times this may be quite simple from an ecological point of view. In salt marsh restoration, for example, the restorationist, having attended to the grade and elevation of a site, may introduce only one or two species of dominant grasses, allowing the numerous other organisms of the salt marsh community to find their way in and sort themselves out on their own.

This may be effective. It may also be efficient. And it is obviously the way restoration must be done in many cases, as Jared Diamond points out in Chapter 23. It is also obvious, however, that the heuristic value of such a process is severely limited. A lot goes on in the healing of a salt marsh in this way that the practicing restorationist does not control and may not even be aware of. This, however, is of interest not only to the "theorist", but to the practising restorationist as well, since not knowing what is going on in a system limits his or her ability to deal with it under varying conditions. As restorationists know, what works on one piece of land this year may fail miserably on another piece next year, and both the theorist and the practitioner want to know why. This, then, argues strongly for the value of restoration efforts conceived and carried out specifically as experiments designed to reveal how the system works. This implies replicas, controls, and clearly defined hypotheses. It may also imply aiming for a smaller, more precisely defined target than usual. Here, for example, the objective will be not simply creation of "salt marsh" or "prairie", but the creation of a community with certain predetermined characteristics – of species composition, nutrient cycling, and so forth.

In short, it may be that restoration has heuristic value insofar as it implies the ability to define goals with precision and to achieve them with confidence. The essential idea is control – the ability not only to restore quickly, but to restore at will, controlling speed, decelerating change as well as accelerating it, reversing it, altering its course, *steering* it, even preventing it entirely (which of course is actually a frequent objective of the ecological manager).

This in turn leads to an important distinction between copying and imitating. Restoration in the narrower sense implies an item-for-item reproduction of the system. But it is just this kind of management that may actually be *least* likely to provide a critical test of understanding. So long as the restorationist copies nature in detail, concretely and literally, his or her aims converge with the natural tendencies of the system. Nature itself then pulls the project towards success, reducing its heuristic value.

What is needed, therefore, is not rote copying, but imitation – the distinction being that copying implies reproducing systems item for item, while imitation implies creating systems that are not identical but that are *similar* in critical ways and that therefore *act* the same. It is imitation, then, and not copying, that is the critical test of understanding, because it is this that implies reproduction of the essentials, the critical parameters of the system grasped as abstractions.

Specifically, this means not just matching community to site, but actually being able to create the same community under various conditions. It also means being able to create communities that resemble other communities in various ways, but that actually *differ* from them in species composition.

It is in this sense, we think, that restoration offers critical tests of ecological ideas and a way of getting beyond the variety of ecological phenomena to discern the essential similarities between them. Ultimately, it is in the identification of resemblances that a science reaches maturity; and it is also in this way that it offers back to practice the generalizations that are the practitioner's most powerful tool.

With these considerations in mind, we may now proceed to a brief introduction to the 26 essays brought together in this volume. The entire book is organized around the idea of restoration as a heuristic process, and amounts to a discussion of this idea from a variety of points of view (Fig. 1.3). Basically, then, it is not a book about results so much as it is a book about a technique – a way of doing research. For this reason, we have

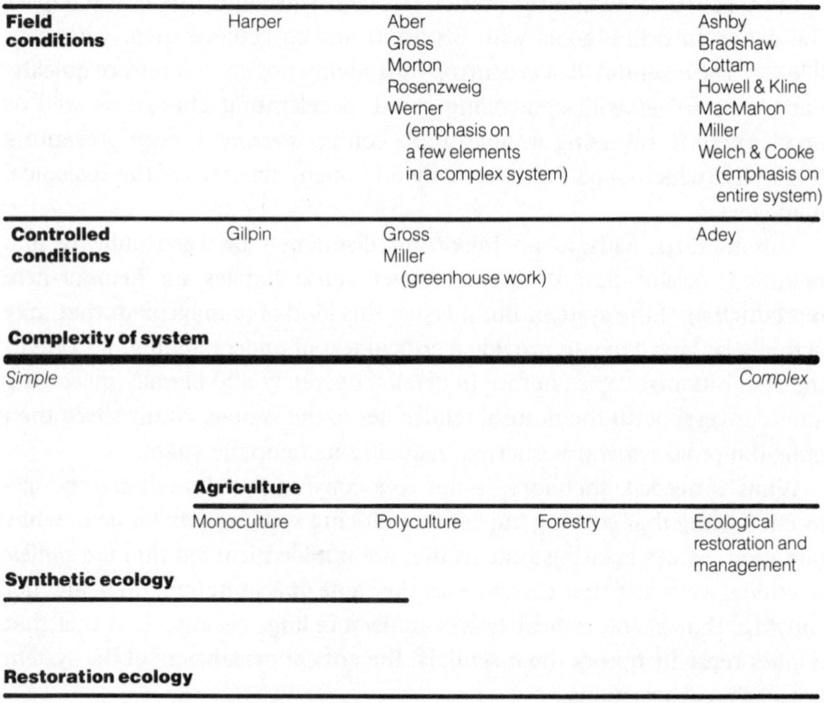

| **Field conditions** | Harper | Aber<br>Gross<br>Morton<br>Rosenzweig<br>Werner<br>(emphasis on<br>a few elements<br>in a complex system) | | Ashby<br>Bradshaw<br>Cottam<br>Howell & Kline<br>MacMahon<br>Miller<br>Welch & Cooke<br>(emphasis on<br>entire system) |
|---|---|---|---|---|
| **Controlled conditions** | Gilpin | Gross<br>Miller<br>(greenhouse work) | | Adey |

**Complexity of system**

*Simple*                                                                                                    *Complex*

**Agriculture**

| Monoculture | Polyculture | Forestry | Ecological<br>restoration and<br>management |
|---|---|---|---|

**Synthetic ecology**

**Restoration ecology**

*Fig. 1.3.* Restoration ecology implies an experimental, synthetic approach to research extending along a continuum, ranging from work with relatively simple systems (left) to more complex, naturally occurring systems (right); and also carried out either in the field (upper section) or under controlled conditions in the laboratory (second section down). This continuum also indicates the close relationship between ecology and agriculture and agricultural research, emphasized by John Harper (Chapter 3). As a form of agriculture that deals with increasingly complex – and "realistic" – communities, restoration ecology promises to extend the value of this approach to the naturally-occurring systems that are of most interest to ecologists. Mapping of representative chapters of this book onto this continuum suggests the range of applicability of the technique of restoration ecology, and also gives a rough idea of the organization of the book.

deliberately sought researchers with a wide variety of experiences and points of view. As a result the book is extremely eclectic. At one level, it may appear eclectic to the point of incoherence. But there is definitely a theme. Each author is represented here because he or she has in some way carried out, participated in, or perhaps simply taken advantage of, an ecological restoration or some kind of effort to construct an ecological system and is in a position to discuss the heuristic value of that effort, and its implications for the practice of restoration.

Throughout, the purpose has been to explore, and as far as possible, to evaluate the synthetic approach to ecological research. The idea has been to discover just how attempts to create systems (whether in the field or in the laboratory, whether as wholes or as parts) have contributed to ecology, and how this approach might be more fully and systematically developed. To a considerable extent, this has meant departing from the norms of scientific writing and abandoning the impersonal and expository in favor of the personal and the narrative. We have encouraged our contributors to do this, and we very much appreciate their cooperation and intellectual generosity in taking the trouble to do this and to undertake the difficult task of looking at a familiar subject in a novel way. We hope our readers will appreciate this as well.

The chapters themselves represent work with the synthesis of very different kinds of ecological systems and also very different kinds of relationships to the process of synthesis itself. Most easily recognized as "restoration" in the usual sense of that word will be the four chapters of Section II, which summarize the experiences of people who have done extensive research on the restoration of representative, "real" communities – prairies, forests and lakes – in the field.

Since one of our objectives, however, is to explore the possibility of linking work of this kind with the tradition of synthetic research in ecology, we then shift "down" in Section III to the domain of what we have called *synthetic ecology*, to consider the work of two scientists who have assembled systems in the laboratory for explicitly heuristic purposes.

The next section (IV) deals with a form of research we see as linking these two extreme forms. This is what we have referred to as *partial restoration*, meaning by this research carried out in the field and with whole systems, but concentrating on the replacement or manipulation of one or a few elements in an attempt to answer specific questions about them and about their role in the community.

The Vth section then returns to examples of actual restoration efforts that have not involved the deliberate synthesis of systems for explicitly experimental purposes, but that have nevertheless taken advantage of restorations set up for other, more or less "practical" purposes. Two of these examples come from the UW Arboretum. A third describes the value of profoundly disturbed areas for studies of evolution.

In the final section, several ecologists reflect on possibilities for actually doing restoration ecology deliberately, by taking advantage of disturbed areas, and setting up restorations specifically as experiments. It is here, perhaps, that the conception of restoration ecology emerges most clearly – restoration as both environmental technology and ecological technique, the comprehensive form of medicine, a science *and* art of healing at the community and ecosystem level.

Overall, although the authors approach the subject in various ways and from different directions, certain ideas clearly recur in virtually every chapter, and it is possible to discern the outlines of a pattern in the results. Two examples seem especially worthy of mention here.

The first is the value of synthesis as a way of determining the ecological importance of phenomena that may have been identified in the course of descriptive or analytical studies, but the significance of which had not been fully appreciated. Again, a conspicuous example is the research on the ecological significance of prairie fire that has been carried out in the course of prairie restoration efforts. But numerous other examples appear in the chapters that follow. Examples are the recognition of the importance of silt and detritus as factors in the eutrophication of lakes (Ch. 8), the role of mycorrhizae in the development of plant communities (Ch. 14), or the effects of circulation of ocean water in the ecology of a coral reef (Ch. 9). In all these instances, the occurrence of the phenomenon was known, but restoration efforts led to a clearer understanding of its ecological importance.

A second theme is that of clearer definition and discrimination of ideas, the sharpening of a concept as a result of experiences with restoration. An example is the splitting of the concept of "mesic" conditions into two components, described by John Aber in Chapter 16.

Of course, numerous questions occur or are implicit throughout the discussions that follow. Among the most important of these is the question of the relationship between the rigorous synthetic work in the laboratory and the work of "real" restoration in the field. For example, just what is the relationship between, say, Mike Gilpin's assembly of communities of fruit flies in milk bottles (Ch. 10) and the experimental bringing together of prairie plants proposed by Virginia Kline and Evelyn Howell in Chapter 6?

Another question pertains to the relationship between the synthetic effort and what has been learned as a result of this effort. Even granting that much has been learned as a result of restoration efforts, was the restoration crucial to the process of discovery, or merely a convenience? No doubt the importance of fire in prairies would have been discovered even if restoration efforts had not been undertaken. But what about the other insights discussed in subsequent chapters?

Related to this is an even broader question: to what extent will it prove fruitful to undertake restoration projects conceived at least partly as experiments to test basic ecological ideas?

These are questions we cannot hope to answer conclusively here, but we find them intriguing, and we hope the readers of this book will find them to be so as well.

### References

Aber, J. D. & Jordan, W. R. III (1985). Restoration ecology: an environmental middle ground. *Bioscience*, **35** (7), 399.

Anderson, R. C., Loucks, O. L. & Swain, A. M. (1969). Herbaceous response to canopy cover, light intensity and throughfall precipitation in coniferous forests. *Ecology*, **50**, 255–63.

Bradshaw, A. D. (1983). The reconstruction of ecosystems. *Journal of Applied Ecology*, **20**, 1–17.

Collingwood, R. G. (1946). *The Idea of History*. Oxford: Oxford University Press.

Curtis, J. T. (1959). *The Vegetation of Wisconsin*, p. 475. Madison: University of Wisconsin Press.

Curtis, J. T. & Partch, M. L. (1948). Effect of fire on the competition between blue grass and certain prairie plants. *The American Midland Naturalist*, **39** (2), 437–43.

Darwin, C. (1859). *The Origin of Species*, pp. 55, 57, 58. New York: Modern Library.

Harper, J. L. (1982). After Description. In *The Plant Community as a Working Mechanism*, ed. E. I. Newman. Special publication no. 1 of the British Ecological Society. Oxford: Blackwell.

Leopold, A. (1934) The Arboretum and the University. *Parks and Recreation*, **18** (xviii), 59–60.

McNab, J. (1983). Wildlife management as scientific experimentation. *The Wildlife Society Bulletin*, **11**, 397–401.

Odum, E. (1984). The mesocosm. *BioScience*, **34** (9), 558–62.

Van Rooyen, D. J. (1973). Organic carbon and nitrogen states in two hapludalfs under prairie and deciduous forest as related to moisture regime, some morphological features and to response to manipulation of cover. PhD Thesis, The University of Wisconsin-Madison.

Wilson, E. O. (1978). *On Human Nature*, p. 18. Cambridge: Harvard University Press.

## A. D. Bradshaw

Department of Botany, University of Liverpool

# 2 | Restoration: an acid test for ecology

In his introduction to *A Sand County Almanac*, Aldo Leopold (1949) wrote, "Conservation is getting nowhere because it is incompatible with our Abrahamic concept of land". In the 1940s, when that was written, plant ecology was still a second-rank subject in his country, as it was in many others. Across the Atlantic it had already captured the attention of some powerful minds, such as that of Tansley, who had not only given us the word ecosystem but had ten years previously published the remarkable document *The British Isles and Their Vegetation* (Tansley, 1939). This not only recorded the complete range and variety of British vegetation, but also gave a penetrating analysis of the factors that determined it. Nevertheless, even in Britain, "ecology" was not the household word it has since become.

So, although it was possible at that time to "see land as a community", the ancient view of land as a commodity was, as Leopold argued, deeply ingrained in our culture. We all know the heritage of degraded and derelict land that these attitudes have produced in some countries. In 1974, in the United States, the total area of substantially degraded land resulting from surface mining alone was over 1 784 000 ha. In the United Kingdom the total amount of officially recognized derelict land was over 43 000 ha. But these figures by no means represent all the damage done. Some estimates suggest that the total areas of degraded land are five times this. We are all also aware of the recent awakening of concern over land, which in both countries has resulted in legislation to ensure that the destruction of land and its associated ecological communities for mining and other purposes is followed by proper restoration. The notable steps have been the *Federal Surface Mining Control and Reclamation Act* of 1977 in the United States, and the *Town and Country Planning Act* of 1971 and the *Minerals Act* of 1981 in the United Kingdom. Other countries have produced similar legislation.

Nevertheless all is not well. We are discovering that the new awareness and the new legislation are not yet stemming the tide of destruction. Both the inexorable rise in world population and the pace of technological change mean that destruction continues to outstrip restoration or natural recovery processes at an alarming rate. All types of ecosystems are being destroyed. The extent of the problem is shown by careful government surveys carried out county by county in the UK in 1974 and 1982 (Department of the Environment, 1984). The amount of derelict land increased over this period from 43 273 ha to 45 683 ha, even though 16 952 ha were reclaimed as a result of major government-financed derelict land reclamation programs. The results are, no doubt, a fairly accurate indication of the predicament of most developed countries and will certainly apply soon to developing countries if they do not take the necessary preventative measures in time. This is not an encouraging situation: at least Alice and the Red Queen managed to run fast enough to stay in the same place. Volumes could be written on the causes, as well as on the shattered hopes and aspirations that lie behind these figures. A more detailed discussion of this situation is given by Bradshaw (1984).

But industrial and social changes are only half the problem. The other half has been the contribution of scientists, or to be more precise, the lack of it. Until recently ecologists have been all too little involved in the restoration of derelict land. The reasons for this relate more to the academic nature of ecology than anything else. Ecology was originally perceived essentially as a descriptive exercise. It was only the contributions of ecologists such as Tansley, Ellenberg, Curtis and Billings that turned it into an analytic exercise, and it is only recently that it has become properly synthetic, despite the holistic views of early workers such as Clements.

But this synthetic approach has been theoretical, not practical. Ecologists have put their efforts into reconstructing ecosystems on paper. People such as Odum, Whittaker, and May, as well as many of the recent contributors to the International Biological Program such as Perkins, Duvigneaud and Clark, have all tended to give us descriptions.

Again, it is profitable to go back to the all too brief writings of Aldo Leopold. He, too, was essentially a great describer. His *Almanac* is a meticulous description of the many facets of nature. Yet there is also contained within it the quite remarkable *Odyssey*, written for the *Audubon Magazine* in 1948. Especially considering the date when it was written, this short essay gives a quite remarkable synthetic insight into the functioning of ecosystems, as well as the contrast between a healthy and a degraded one.

But Leopold went further than just putting down his synthetic views on paper. He became a tireless proponent of practical conservation, and on an abandoned sand farm in Wisconsin tried to rebuild "with shovel and axe" what was being lost elsewhere. He became one of the first of the new ecologists, who not only saw the need for restoration of our battered inheritance but also practised it. This was the great step forward – not just to talk and write about ecology, but to test out ideas in practice.

A history of restoration ecology has still to be written. However, it is significant that, before Leopold, any practical work on ecosystems – developing or manipulating them – had been in effect the business of agriculture, range management, and forestry. But perhaps even worse, the reconstruction of ecosystems had been almost entirely the business of civil engineers working on sites such as road cuts, spoil banks, and sea defenses. Ecologists had been conspicuous by their absence. Because of this, they have missed the critical opportunity not only to make practical contributions to our environment, but also to test out their fundamental ecological ideas in practice. The latter point is crucial, for practice is a stern master and a test of our competence.

### Practice as an acid test

Let us begin with a brief look at the history of our understanding of ecosystems. Every ecologist will agree that ecology has been dominated in the past by two approaches: the organismic approach of Clements, which implies that an ecosystem is more than the sum of its parts; and the individualistic approach of Gleason, which implies that an ecosystem is really only the simple product of the behavior of its constituent species. Despite the recent wide adoption of a reductionist analytical approach, McIntosh (1985), in his recent excellent review, argues that much of the spirit of the organismic approach persists. Perhaps what is patently obvious is simply that there are a large number of theories about the nature of ecosystems – and more particularly about the nature of succession (McIntosh, 1980).

Theories are essential for intellectual progress, but if they become divorced from reality they may become no more than an intellectual game. Similarly, in the process of analysis, although certain characteristics may be held to be critical, it is difficult for the analytical approach alone to confirm such conclusions. The evolutionary zoologist A. J. Cain (1977) wrote, "The golden rule is always to ask questions of the animals, not the pundits. However vociferously any particular theory...may be proclaimed, it is still necessary to ask whether it actually applies to any real populations". In the same way, Peter Medawar (1967) has written, "Scientists are building explanatory structure, telling stories which are scrupulously tested to see if they are stories about real life". This has crucial implications for ecology and ecological restoration, because for any natural phenomenon there is only one ultimate test of our understanding. That is to see if our ideas actually work in practice. Restoration provides such an acid test because each time we undertake restoration we are seeing whether, in the light of our knowledge, we can recreate ecosystems that function, and function properly. This is a particular version of the Baconian idea of verification by experimentation.

An analogy that will make this clear is provided by any piece of machinery, such as a watch. It is easy to take a watch apart and to describe its parts. From their structure we can even infer their function and can feel we understand how the watch works, and indeed how it can be made to work. Armed with this understanding we can put the watch together and see if we can make it keep correct time. If we succeed, we have tested our understanding quite critically. But it would be a more critical test of our understanding if we were given a watch that had been trampled on, or a watch from which some of the pieces had been lost. If we could, from either of these starting points, repair the watch so that it again kept perfect time, we would certainly have demonstrated an understanding of the way watches work.

The essential element of proof of understanding by the restoration of function is that, because the object being reconstructed is complex, failure to reconstruct any part of it properly will cause it to function improperly or fail to function at all. And, generally, the more complex the object, the more stringent this test of understanding. Engineers will be very familiar with this idea of a close relationship between understanding and the ability to make or restore. Yet it seems to have escaped the attention of philosophers, even those like Bertrand Russell (1948) who are interested in the nature of human knowledge, perhaps because philosophical thinking has been dominated by the analytical approaches of most scientists.

Nevertheless, from these arguments we can logically argue that the successful restoration of a disturbed ecosystem is an *acid test* of our understanding of that system. In other words, there can be no more direct test of our understanding of the functioning of ecosystems than when we put back, in proper form or amount, all the components of the ecosystem we infer to be crucial, and then find that we have recreated an ecosystem that is indistinguishable in both structure and function from the original ecosystem, or the ecosystem that served as our model.

In this reconstruction we can pay attention to whatever theories we think are important, and incorporate what individual components or functions we consider to be critical. If, however, we leave out anything that is critical because our understanding is incomplete or in error, then the ecosystem will function improperly. It is this procedure for testing and refining ecological ideas that is the subject of this book, and numerous examples of this process will be described in some detail in subsequent chapters.

These accounts certainly support the logic suggesting the value of restoration as a test of ecological understanding. At the same time, there are two reasons why I believe it cannot constitute an absolutely critical test. The first reason is that given by Popper (1980) in his famous work on the logic of discovery. Unfortunately it is not logically possible to prove that something is correct, because there is always a possibility that some alternative correct explanation exists. The only logically certain operation is refutation. Proof is unfortunately asymmetrical. In the context of restoration, this of course means that we will learn more from our failures than from our successes since a failure clearly reveals the inadequacies in an idea, while a success can only corroborate and support, and can never absolutely confirm, an assertion.

The second reason is more specific to ecology, and relates to the nature of the material with which ecologists are dealing. This material is not inanimate and has considerable capabilities of self-repair. Indeed, given sufficient time, almost any degraded or destroyed ecosystem will restore itself. This self-healing capacity of ecosystems is an important and valuable attribute. But it does limit the value of restoration as an acid test of understanding because is means that a successful restoration could result from these self-healing properties despite faulty theory and incorrect restoration treatments.

This objection is troublesome because it does appear to reduce the stringency of restoration as a test of our understanding. Yet put in a sensible context it is not a serious problem. What we must include is a criterion of the *rate of achievement* of the successful restoration. In the absence of any treatment, self-restoration of almost all ecosystems is relatively slow. So when we are considering successful restoration as an acid test, we must imply restoration within a short period of time, a period much shorter than that which would be necessary for natural processes. This extra criterion is easy to apply in nearly all situations. Indeed, in the case of restoration work on degraded or derelict land it is usually a prime practical consideration as well as a theoretical one.

## The value of restoration practice

The basic principles of land and ecosystem restoration are the same as the basic principles of ecological succession, although we must remember that in many situations we may be dealing with primary rather than secondary succession. In both, we are interested in what determines the development of ecosystems from very skeletal beginnings and what may restrict it.

The essential quality of restoration is, however, that it is an attempt to overcome artificially the factors that we consider will restrict ecosystem development. This gives us a powerful opportunity to test out in practice our understanding of ecosystem development and functioning. The actual restoration operations will often be dominated by engineering or financial considerations, but their underlying logic must be ecological.

Ecosystem restoration therefore has a duality. On the one hand, it is merely a technical business, often with an ecologically crude objective, such as the establishment of some sort of permanent vegetation cover as quickly and cheaply as possible. Even this, however, may not really be quite as simple as all that. Speed means getting everything right, which in turn means having a real understanding of the factors limiting plant growth and not missing out on any of them. Cheapness also depends on understanding, because obviously if we understand the natural processes we are working with well enough to harness and take advantage of them, we can work more efficiently. Furthermore, long term maintenance costs will be reduced if we produce a product that functions correctly (i.e., normally). And to do this we need yet further understanding about how a stable, mature ecosystem works and maintains itself.

This brings us to the other half of the duality. The process that may seem to be just a technical operation actually provides a unique, fundamental challenge to our understanding of ecosystems. In doing this, it provides us with tests that are not vague and holistic, but are precise appraisals of individual aspects of ecosystem function. These appraisals essentially test everything, because if properly carried out and interpreted, they will show not only what we know but also what we have forgotten. And they represent an important way of putting ecological research into context, because they demonstrate which functions are of crucial importance, which are only of subsidiary importance, and even how they all interact. These appraisals are therefore both reductionist and synthetic.

It would be difficult to find any better test for our developing theories and understanding. We should surely, therefore, look very carefully at all that we can glean and conclude from ecological restoration work. This is something that we have, so far, hardly begun to do.

### References

Bradshaw, A. D. (1984). Land restoration: now and in the future. *Proceedings of Royal Society*, London, B223, 1–23.
Cain, A. J. (1977). The efficacy of natural selection in wild populations. In *Changing Scenes in Natural Sciences 1776–1976*, ed. C. E. Goulden, pp. 111–33. Philadelphia: Academy of Natural Sciences.
Department of the Environment (1984). *Survey of Derelict Land in England 1982.* London.
Leopold, A. (1949). *A Sand County Almanac.* New York: Oxford University Press.
Medawar, P. B. (1967). *The Art of the Soluble.* London: Methuen.
McIntosh, R. P. (1980). The relationship between succession and the recovery process in ecosystems. In *The Recovery Process in Damaged Ecosystems*, ed. J. Cairns, pp. 11–62. Michigan: Ann Arbor Science.
McIntosh, R. P. (1985). *The Background of Ecology.* Cambridge: Cambridge University Press.
Popper, K. (1980). *The Logic of Scientific Discovery*, 10th edn. London: Hutchinson.
Russell, B. (1948). *Human Knowledge: Its Scope and Limits.* London: Allen & Unwin.
Tansley, A. G. (1939). *The British Islands and Their Vegetation.* London: Cambridge University Press.

*John J. Ewel*

# Restoration is the ultimate test of ecological theory

Ecosystem restoration is an activity at which everyone wins: when successful, we are rewarded by having returned a fragment of the earth's surface to its former state; when we fail, we learn an immense amount about how ecosystems work, provided we are able to determine why the failure occurred.

The most commonly employed criterion for judging the success of an ecosystem reconstruction project is simply whether or not the reconstituted community resembles the original: does it contain the same dominant species and have similar physiognomy? But such superficial comparisons often prove deceptive when, in the longer term, the recreated community disintegrates. The success of ecosystem restoration can be judged by the five criteria described below. The ecologist capable of creating an ecosystem that passes this rigorous test earns high marks; the one who fails is sure to gain new insights into ecosystem structure and function.

### Sustainability

Is the reconstructed community capable of perpetuating itself, or, like agricultural ecosystems and golf courses, can it be sustained only if managed by people? Germination and ecesis are the most precarious phases of plant-community development, but these stages can be bypassed during restoration by planting seedlings rather than seeds. The failure of the community to regenerate thereafter means either that the environment changed, that the restored community was a seral stage, or that the ecologist did not understand the regeneration requirements of the species.

### Invasibility

Does the reconstruction yield a community that resists invasions by new species? Intact, natural communities are, in general, less easily invaded than ones that have been damaged or ones that lack one or more of their key species. Invasions can be symptoms of incomplete use of light, water, and nutrients.

### Productivity

Like invasibility, productivity is dependent upon efficacy of resource use by the community. A restored community should be as productive as the original. Net ecosystem productivity is an especially useful measure of community performance because it integrates many processes, including photosynthesis, respiration, herbivory, and death.

### Nutrient retention

Although all ecosystems are open to nutrient fluxes, some are more open than others. A reconstructed community that loses greater amounts of nutrients than the original is a defective imitation. In the long run it will prove to be unsustainable because it will be invaded by new species and its productivity will decline.

### Biotic interactions

Reassembly of formerly associated plant populations often – but not always – leads to reconstitution of the entire community. Animals and microbes usually colonize spontaneously because of their mobility and ubiquity, respectively. Nevertheless, biology texts are packed with examples of communities whose functional integrity hinges on a pollinator, a microbe essential for nitrogen fixation or phosphorus uptake, or a key link in a food chain. The importance of key species is often best revealed by their absence.

Ecologists have learned much about ecosystem structure and function by dissecting communities and examining their parts and processes. The true test of our understanding of how ecosystems work, however, is our ability to recreate them.

*John L. Harper*

(Unit of Plant Population Biology), University College of North Wales

# 3 | The heuristic value of ecological restoration

The *raison d'etre* for a science of ecology is presumably the development of an understanding of the workings of nature that would enable us to predict its behavior and to manage and control (conserve or change) it to our liking. It is very doubtful, however, whether much that has so far composed the science of ecology has actually had these aims clearly in mind. This has been especially true in plant ecology, where research has been dominated by description and correlation, which are only the preliminaries to posing and answering questions about causation (Harper, 1984).

Of course, description and correlation do have predictive value. For example, once we have found that there is a correlation between the distributions and abundance of species and the temperature, water regime, soil pH and mineral composition of a site, we can very often predict the composition of vegetation on the site if we know these factors. Conversely, if we have information about the vegetation we can often draw conclusions about conditions of the site with some assurance. In the same way, accurate descriptions of vegetational successions in the past may enable us to make reasonably accurate predictions about the nature of future successions. This approach, however, is severely limited and rarely if ever provides the basis for generalizations that would allow us to make predictions for other, ecologically dissimilar areas. What this tells us, of course, is that the descriptions and correlations only hint at the causal factors that underlie them. And indeed the current arguments about the nature of the forces that determine successions (Connell & Slatyer, 1977) or about the role of competition between organisms in determining community composition (Schoener, 1983; Connell, 1983) show just how dimly we understand ecological causation.

It is this state of affairs that argues for the importance of a more manipulative, experimental approach to ecological research such as that represented by restoration ecology. I have argued elsewhere (Harper, 1984) that there is an analogy between the study of ecological phenomena and the study of clocks and watches, and this analogy is useful here. If, for example, we were to sample wristwatches collected from a human community, we might be able to classify them or to arrange them in some kind of order. They may have hands that move or figures that flash. They may be made of steel, silver or gold; the strap may be leather or metal. We would probably also be able to establish a loose correlation between these characteristics and the age, sex and wealth of the owner. If a watch were lost, this information would be useful in locating its owner, but it would be of little use if the watch were broken and the owner wanted to repair it or make a new one.

The repairer of a watch does not need a description or a set of correlations, but a kit of parts and the knowledge of how to fit them together. With this kit and this knowledge, the repairer might hope to reassemble the watch, and if this is done properly the watch will have acquired the emergent property of an integrated whole, measuring and reporting the passage of time. To discover how a watch works, the repairer does not simply describe it, but takes it to pieces and puts it back together and, if it works, gains at least some understanding of the way in which it does so. The point is that this understanding is achieved by *doing* something to the watch in an attempt to get it going again. And this approach may be useful even if rather crudely applied. For example, an old remedy for a watch that stops is to boil it! If it then starts to function again it is likely that oil or some other soluble material had clogged the works. And of course, if one can learn something

*Fig. 3.1.*    Much of agriculture and forestry involves the deliberate creation of monocultures of a desired species and also their maintenance and protection from other pest species such as weeds and disease organisms. The very creation of such monocultures involves heuristic exercises in ecology (photograph by B-Wolfgang Hoffmann).

by trying to repair a watch that has been broken accidentally, one might learn even more by perturbing it deliberately and perhaps then trying to repair it. This often proves a valuable way to determine exactly what is crucial in the operation of any system and why it is crucial. In the case of a watch, for example, removing the strap will normally leave it with most of its time-keeping properties intact, whereas removing a spring or a battery quickly reveals a crucial part of the system. Later in this volume (Ch. 13), Mike Rosenzweig discusses the importance of a similar approach to the study of ecological communities.

The crucial point here is that to repair a watch it is not enough to know what it looks like, however precisely. It is necessary to understand how it works, to know cause and effect, and to understand which cogs drive which cogs, how power is transmitted, which parts wear out more quickly, and so forth. This is especially true if one is going to try to repair different *kinds* of watches under varying conditions. And of course, the same thing is true of an ecologist or restorationist working with communities or ecosystems, no two of which are ever exactly alike. To restore such systems successfully he or she must understand how they work. Conversely, perturbing the system, taking it apart and trying to reassemble it, is one of the best ways to achieve this understanding. Restoration ecology is just this process of assembly, carried out specifically to test ideas about how communities and ecosystems work. Indeed, this approach may well prove crucial to the development of ecology as a mature, experimental and predictive science. If we can create a working ecological system by deliberative assembly of its parts, we can claim to have made a truly predictive ecology. This being the case, it is puzzling that this approach to ecological research has attracted so little attention. Astronomers, after all, would give their souls for the chance to study the universe in this way – even to change the position of a single star!

Of course some will argue that natural ecological systems have holist properties that represent more than the sum of their parts plus their inter-actions. If this is so, the study of these systems should perhaps be carried out by theologians rather than scientists. A holist view of ecological systems prevails in many quarters, and this is especially evident in the way in which ecology is commonly taught. It is perhaps the only science that is typically presented to students as a descriptive exercise, rather than a search for cause and effect that begins with the simplest systems and proceeds step by step to explore those that are more complex. "There is probably no other science in which students are taught by being dropped into the deep end of complexity. Ecology is usually introduced to school children by showing them oak woods, chalk grassland, an intertidal zone, or a pond or lake margin, and perhaps asking them how to describe it. This is equivalent to introducing a child to chemistry by showing the structure of hemoglobin or a DNA helix" (Harper, 1982). This may be a good way of showing the student the immensity of the questions facing the ecologist. But the next stage in teaching ecology *as a science* needs to be the study of the very simplest ecological systems that can be assembled (a population of one species of grass in a tray of soil, for example, or a single species of alga in a culture solution). This is rarely done, either in high school or at universities.

Of course, what is true in the schools is also true in practising the science. If we accept (as an act of faith in the methods of science) that "the activities of communities of organisms are no more than the sum of the activities of their parts *plus* their interactions, it becomes appropriate to break down the whole into the parts and study them separately. Subsequently, it should (again as an act of faith) be possible to reassemble the whole, stage by stage, and approach an understanding of its workings. It must mean looking at individual organisms and their behavior alone and when brought together in the very simplest communities of single species in very simple physical conditions. From this stage increasing complexity can be introduced step by step" (Harper, 1982). This is exactly the procedure of restoration ecology which, therefore, is central to the logic of genuine ecological inquiry, not marginal to it or a spin-off from it. Nevertheless, most of the research that is relevant to this approach has been done by agricultural researchers and foresters – not by those calling themselves ecologists!

## Relevance of agriculture and forestry to ecological research

### Monocultures

Much of agriculture and forestry involves the deliberate creation of monocultures of a desired species and also their maintenance and pro-tection from other pest species such as weeds and disease organisms. The very creation of such monocultures involves heuristic exercises in ecology, most notably in the choice of species and genotypes, and in the regulation and control of interspecific competition.

There are probably no crops, forests or livestock that are economically grown at densities so low that individuals do not interfere with and reduce the growth of their neighbors by making demands on limited resources. The search for optimal stocking and cropping densities, therefore, has provided some of the most elegant insights into the nature of intraspecific competition. I have already reviewed a number of examples from crops and forests (Harper, 1977), and there are numerous examples from animal husbandry as well. Indeed, although most ecology text books cite laboratory studies of *Drosophila*, *Paramecium* or *Tribolium* in discussions of density-dependent processes, much of the best research in this area has been carried out in the area of livestock and poultry husbandry.

Such studies, moreover, may be directly relevant to interpretations of more complex natural ecological systems. The establishment of mono-cultures in agriculture and forestry, for example, has already led to an ecological generalization of great importance in our understanding of the importance of diversity in natural communities. This is the recognition that genetic variation within a population may hinder or even prevent the development of epidemic outbreaks of pests and diseases. There is now good evidence from agricultural practice that epidemic outbreaks of diseases such as *Phytophthora infestans* (late blight of potatoes) and *Puccinia* spp. (stem and leaf rusts of cereals) may be prevented if the monoculture is made genetically diverse (Burdon & Shattock, 1980). Such findings obviously have important implications for natural ecological communities.

### Mixtures of species

Agricultural and forestry research have also produced a large body of literature about the interactions that occur between populations of two or more species, especially between crops and weeds. An especially striking example is the study of the ecology of skeleton weed (*Chondrilla* spp.) in Australian cereal crops. Skeleton weed is an introduced plant that thrives within cereal crops (themselves introduced species), and may seriously reduce cereal yield. In an attempt to control skeleton weed, a rust (*Puccinia chondrillina*) was introduced to Australian grain fields in 1971. This effort was partially successful, and wheat yields increased when skeleton weed was suppressed, showing that the weed was indeed competing with the wheat. Interestingly, however, the disease proved aggressive against only part of the weed population, leading to the discovery that the weed popu-lations were composed of several distinct genotypes (apomictic races) that differed in resistance to the rust. Where the disease was introduced sus-ceptible forms of the host were severely attacked, but the more resistant forms then increased in abundance, implying that there had been intraspecific competition within the skeleton weed populations (Burdon & Shattock,

*Similar indirect effects to A. epos L.th system.*

1980). This study involved an entirely synthetic system – wheat, three apomictic races of skeleton weed and the rust. Yet it provided information pertaining to succession, the role of disease in shaping a community, intra- and interspecific competition and the influence of genetic variation on the structure and stability of a community. And this is not an unusual example. Comparable studies are abundant in the literature, many of them involving biological control, in which the balance between two species in a community (typically a crop and a pest) has been altered experimentally by additional species introduced to control the pest.

It is of special interest here that biological control programs in agriculture usually aim not to eliminate a pest but to reduce it to economically less damaging levels. This being the case, it is important that the introduced agent becomes established as a permanent part of the community. For this reason, if for no other, such projects provide remarkable opportunities to study how natural communities work, and should be regarded as providing model experimental systems for the student of restoration ecology.

Although interest in the multi-species gardens characteristic of certain aboriginal forms of agriculture has increased in recent years, the deliberate combining of two or more crop plant species is not common in modern agriculture. In sown pasture systems it is common for two or more species of grass and one or more species of legume to be sown together. The mixture may be self-sustaining for its nitrogen resources because of the presence of the legume, productive because of the vigor of grass growth, and nutritious because of the higher protein content of a grass-legume mixture than of grass alone. A major problem in agronomy has been how to achieve and maintain the ideal balance of species in such a sown grassland. A highly sophisticated technology has developed to this end. "Studies that developed particularly from the work of the pioneer agronomists (e.g., Jones, 1933) led to the remarkable situation in the 1950s in which various farming organizations in the UK were giving medals, silver cups or prizes to farmers who had achieved 'ideal' pasture composition. Careful balancing of nitrogen and phosphorus fertilizers and controlled grazing can produce a desired balance of species. There is probably no example among world vegetation types of a community so deeply understood and with its interactions so deeply researched" (Harper, 1982). In some environments, especially on acid soils, the seed of the legume introduced into such sown grassland is deliberately infected before sowing with appropriate, highly efficient nitrogen–fixing strains of *Rhizobium*. Recently, seed producers have introduced white clover (*Trifolium repens*) seed in pellets that include propagules of *Rhizobium* and mycorrhizal fungi. What the farmer sows and manages is then a carefully designed mixture of plant species, complete with the makings of its own rhizosphere flora. This is indeed restoration ecology.

An even more striking example of the deliberate management of a community of two or more species comes from the developing practice of agroforestry. Forestry plantations have traditionally been monocultures. But agriculture and forestry can be integrated to create a patchwork of woodland interspersed with patches of grassland or cropland (a common pattern in Finland). More dramatic, and even more closely related to the restoration of native ecological communities, is the establishment of forest trees within a grazed grassland. This practice, which is developing especially rapidly in New Zealand, involves planting Monterey pine (*Pinus radiata*) at lower than normal forest densities in a grassland grazed by livestock. The trees are protected from the livestock for the first few years, after which livestock is permitted to graze beneath the developing tree canopy. Gradually, the density of livestock that can be supported declines as the system changes from grassland to forest. The trees are then harvested and the cycle is repeated.

A further example from New Zealand agriculture is highly relevant to the theme of restoration ecology. There has been a growth of interest in the use of a variety of livestock as a means of managing grazed pastures and maximizing their productivity. Many grassland areas in New Zealand are dependent on legumes such as white clover for nitrogen. Sheep, however, graze selectively, and can rapidly reduce the clover content of the swards. Goats, in contrast, seem to dislike clover, and clover populations frequently increase in areas grazed by goats. This suggests that mixed populations of sheep and goats may utilize these pastures more efficiently, and also manage them more effectively by maintaining the essential balance between grasses and clover. In this situation, we can see a kind of restoration ecology that creates and maintains a relatively complex community, comprising not only the clover (and its rhizobial nitrogen-fixing microflora) and grasses but also two species of herbivores.

These various examples from agriculture and agroforestry are described here because they clearly demonstrate that the deliberate creation of ecological communities of various degrees of complexity is already being practised and represents a kind of creative or restoration ecology in action. There is, moreover, a vast wealth of information relevant to restoration ecology in the literature of agricultural and forest science, and the restoration ecologist needs to be aware of this if he or she is to avoid "reinventing the wheel".

## A selection of questions that might be answered by the deliberate creation of communities

The science of ecology is full of questions, and many may be approached and perhaps answered by the deliberate design and assembly of communities. Indeed, in some ways the restoration ecologist has greater freedom than the agriculturist or forester in designing and synthesizing experimental communities because he or she is not necessarily required to produce a marketable product. It is important, however, that the restoration ecologist defines his or her questions carefully and plans experiments accordingly, even though this may often be difficult for purely practical reasons. The questions that are listed below are only a few of the vast number to which restoration ecology might give answers. Some are among the most fundamental of ecological questions, illustrating the profound contribution restoration ecology could make to ecological science. Other examples are discussed in detail in subsequent sections of this book.

☐ Does increasing the species diversity of a synthesized community increase its stability and resilience?

☐ How does genetic diversity within component species influence the stability of an ecological community?

☐ If species evolve together, do they tend to become more compatible or less so? There is some evidence that prior experience of selection in each other's presence (forced coevolution) does enhance the ecological combining ability of species (e.g., Evans *et al.*, 1985; Haukioja, 1980; Turkington & Harper, 1979). This is a question that has both evolutionary and ecological implications, and one way to explore it would be to assemble appropriate experimental communities. (See the related discussion by Rosenzweig in Ch. 13.)

☐ How does the age structure of component species influence an ecological community? This question may be especially important for communities comprising long-lived species such as trees. It seems likely that many of the ecological problems that are associated with monocultures, or with populations of species with low genetic diversity, may also be met in communities such as pine or boreal forests, which are dominated by populations with exceptionally even-age structures.

☐ Can succession by bypassed? It is characteristic of areas of land (or water) that are left to be colonized naturally to pass through successional stages. A restorationist may not be content to wait for this process, however, and may indeed want an end product different from that which would occur naturally. There are many theories to account for succession (Connell & Slatyer, 1977; Drury & Nisbet, 1973) which include the ideas that: (a) early species change the habitat in a way that favors later inhabitants; (b) later inhabitants are those that can exclude earlier forms; and (c) succession is mainly a product of the different rates at which dispersal units of the various species arrive at a site. Each of these possibilities represents a hypothesis that can be tested by restoration ecology experiments. If, for example, the answer is (a), the restoration ecologist might speed succession by doing to the habitat what the early colonists might have done. He might, for example, deliberately add organic matter or fertilize with additional fixed nitrogen or some other nutrients that would have been brought to the soil surface by the exploring roots of earlier colonists. If the answer is (b), however, the succession might be hastened by simply arranging for the normally later colonists to be deliberately planted earlier. Finally, hypothesis (c) suggests that the restoration ecologist might hasten succession by ensuring that the species with poor dispersal were deliberately brought in at an early stage. In any case the point is that in carrying out these various procedures the restoration ecologist is testing the validity of these ideas for the community that is being constructed. (See also Chs 6, 7 and 12.)

☐  What role do animals play in succession? There is a strong temptation for the ecologist to assume that the establishment of the vegetation is what comes first and that the animal component will follow naturally, or can be put in almost as an afterthought. However, many plant ecologists are becoming aware that plant communities are what animal communities make them. Restoration ecology offers a way to test the importance of introduced animals in structuring the plant community.

☐  What is the role of mutualists? Some of the most effective early colonists in community succession are mutualists. It is, for example, lichens that are the prime colonists of bare rock. Similarly, legumes and nitrogen–fixing shrubs and trees are effective colonists of mining spoils and exposed subsoils, and mycorrhizal trees seem especially successful on similar sites. This suggests that such miniature communities (tight species complexes) are especially tolerant of environments that have not already been softened by biological activity. Restoration offers ways of exploring this idea.

Ecological science is not sophisticated enough to give general or specific answers to these and similar questions at present. It seems probable that attempts at restoration ecology will be most likely to produce answers. They are all essentially the same sort of questions faced by the watchmaker who is trying to put together the parts that make a working watch. If the attempts of the restorationist to design and create communities to order for practical or environmental purposes are also regarded as attempts to test ideas or answer crucial questions in ecology, a quite new ecological science may emerge to give us the necessary knowledge to take control of our flora and fauna – make it, change it, or conserve it. In this way we will have become like the watchmaker – not just tinkerers but craftsmen and engineers.

### References

Burdon, J. J. & Shattock, R. (1980). Disease in plant communities. In *Applied Biology*, vol. V, ed. T. H. Coaker, pp. 145–219. London: Academic Press.

Connell, J. H. (1983). On the prevalence and relative importance of interspecific competition: evidence from field experiments. *American Naturalist*, **122**, 661–96.

Connell, J. H. & Slatyer, R. O. (1977). Mechanisms of succession in natural communities and their role in community stability and organization. *American Naturalist*, **111**, 1119–44.

Drury, W. H. & Nisbet, I. C. T. (1973). Succession. *Journal of Arnold Arboretum*, **54**, 331–68.

Evans, D. R., Hill, J., Williams, T. A. & Rhodes, I. (1985). Effects of coexistence on the performance of white clover-perennial ryegrass mixtures. *Oecologia*, **66**, 536–9.

Harper, J. L. (1977). *Population Biology of Plants*. London: *Academic Press*.

Harper, J. L. (1982). After description. In *The Plant Community as a Working Mechanism*, ed. E. I. Newman. Oxford: Blackwell Scientific Publications.

Harper, J. L. (1984). Foreword. In *Perspectives on Plant Population Ecology*, ed. R. Dirzo & J. Sarukhan. Massachusetts: Sinauer.

Haukioja, E. (1980). On the role of plant defenses in the fluctuation of herbivore populations. *Oikos*, **35**, 202–13.

Jones, M. G. (1933). Grassland management and its influence on the sward. *Journal of the Royal Agricultural Society*, **94**, 21–41.

Schoener, T. W. (1983). Field experiments on interspecific competition. *American Naturalist*, **122**, 240–85.

Turkington, R. & Harper, J. L. (1979). The growth, distribution and neighbor relationships of *Trifolium repens* in a permanent pasture, IV. Fine-scale biotic differentiation. *Journal of Ecology*, **67**, 245–54.

*Frederick Turner*
School of Arts and Humanities, The University of Texas-Dallas

# 4 | The self-effacing art: restoration as imitation of nature

The reconstruction, species by species, of a damaged or degraded ecosystem is a rather special kind of human activity. In order to do it well, and especially to teach it to others, it might be valuable to have a sound working theory of the human capacities and talents that it involves and calls forth.

What other human activities does it resemble? Science, certainly, but science is normally considered to be a method of analysis, while restoration is necessarily a process of synthesis, of performance. Restoration is indeed guided by analysis. It serves as a test of the analysis it relies on, and is richly suggestive of new analytic models, new hypotheses. But this fertility of hypothesis is called into being by the imperative of successful and positive action. Unlike science, which professes to understand the world by external observation and measurement of it, restoration understands the object of its study from the inside, as the creator – or rather, the recreator – understands the creation.

But is science itself as detached from the world it studies, as purely analytical, as divorced from performance, as its stereotype implies? In quantum physics it is a truism that what the physicist observes is partly the result of the process of observation. The accelerator now creates to be studied particles which are very rare, or even normally absent on this planet. Chemists regularly create polymers that never existed, and molecular biologists produce chimeras stranger than any generated by mutation or sexual recombination. What restoration does is in a sense humbler, but in another sense more miraculous still: to create an ecology that is indistinguishable from the natural reality. How better to understand that reality?

But if restoration is not exactly a science, it is not a *technology* either, in the traditional sense of the word, for its goal is not a product but a process. One could say that the biological machine the restorer produces has no function but its own ordered reproduction. Nor is restoration quite like a *craft*, in which the boundaries are clearly established and the criteria of productivity can be easily measured. In some ways it is more like play, but it is a very serious kind of play, one that submits itself to high and demanding canons of perfection, and which is not embarked on only for its own sake. In fact the activity it resembles most closely is art. Let us call it an art and see where this characterization gets us.

To do this we must know what we mean by art. A simple definition may be best: art is the creation of beauty. But what then is beauty? Here, oddly enough, the perspective of the biologist may be more instructive than that of the esthetician. Recent work in the biological foundations of esthetics, which I have summarized in the final bibliographical chapter of my recent book *Natural Classicism* (Turner, 1986), suggests that the esthetic capacity is an inherited and very sophisticated competence. It is given specific form by culture, to be sure, as is our predisposition to speech, but its deep syntax, so to speak, is wired into our neural hardware. Obviously then it is designed for some very important function, and was selected for and refined by evolutionary mechanisms. It is a capacity we share with other animals, but which is in us enormously more developed, flexible and variable in application.

The new research suggests that the esthetic sense is a capacity to organize and make sense of very large quantities of ill-defined information, to detect and create complex relationships and feedback systems, to take into account multiple contexts and frames of reference, and to perceive harmonies and regularities in them that add up to a deep unity, a unity which generates predictions of the future and can act as a sound basis for future action. The new understanding of the vigorous constructive activity of the sensory cortex suggests that we do not passively receive the world of vision, but actively create it (Marr, 1984). The esthetic capacity is to perception as perception is to mere sensation, or as sensation is to the mechanical effect of some outside event upon an inanimate object. In other words, the unities that the sense of beauty appreciates are a higher, more integrated version of the enduring, solid objects in space and time which it is the job of the senses to construct for our use. Thus a complex unity like an ecosystem is perceived by the esthetic sense as a solid object in space is perceived by sight. The exercise of the esthetic capacity is rewarded and reinforced by the brain's own self-reward system, which we feel as esthetic pleasure.

What capacity could be better adapted to the complex, context-rich work of restoration, which must harmonize into a higher unity large masses of mutually dependent information? Our destiny as a species now appears to be bound up with the success of our attempts to reconstruct our living environment (not only on this planet, perhaps). The sense of beauty tells us what is relevant, what is likely, what is proper, what is fruitful. We would be in a desperate case if the only capacity we could rely on was our logical ability to put two and two together. There is not enough time to work everything out in that fashion, and there is simply too much information and too many possible consequences to do it without those higher integrative abilities.

Aldo Leopold (Callicott, 1984) suggested the possibility of a human natural esthetic. Restoration may now with some justification call upon a human esthetic capacity which is not a merely passive appreciation, and in which the artificial and the natural cannot be distinguished. Indeed, one of our problems is the distinction itself. True gardeners of the planet will not need it any more.

If restoration is an art, what kind of art is it? We have at our disposal – although we have used it very little – a rich storehouse of theory and recorded experience in the field of art and nature. It can be found in renaissance esthetics. Sir Philip Sidney, for instance, defines art as the imitation of nature. By this he does not mean that the artist imitates what nature merely *is*, as a photograph or a diorama imitates the visual externals of a scene, but rather he imitates what nature *does*; that is, he generates a living and self-developing order. Human art, Sidney maintains, can better the current productions of nature, but that is precisely because human art is a natural process. Shakespeare says the same thing in *The Winter's Tale*.

We find similar ideas in other great Renaissance esthetic theorists: the architects Alberti and Palladio; the critic Scaliger; the philosophers Pico, Ficino and Vida; the scientists Bruno, Hariot, and Bacon. Such ideas were the guiding principles of Leonardo da Vinci. In the Enlightenment, they guided the great landscape gardeners of England – Capability Brown for example. We find them in Marvell's *Upon Appleton House*, Pope's *Twickenham*

*Garden,* and in the marvelous scene in which Elizabeth Bennet visits Darcy's country house in Austen's *Pride and Prejudice.*

Now the curious thing is that Renaissance art gave rise to good science; that is, its attempt to imitate nature had profound heuristic value. The discoveries of Leonardo da Vinci in anatomy, aerodynamics, hydrostatics and mechanics, which grew out of his attempt to imitate nature in drawing, painting, sculpture and architecture, are justly famous. Brunelleschi, in the attempt to imitate the natural phenomenon of visual perspective, whose discovery he erroneously attributed to the ancient Romans, invented pictorial perspective himself, and in so doing made large advances in the science of projective geometry. The Fibonacci Series in mathematics was discovered in the attempt to find a way of calculating accurately the Golden Section, which appears frequently in nature and is fundamental to the esthetic laws of proportion. The great Elizabethan scientist Thomas Hariot, a close friend of the poets Christopher Marlowe, Edmund Spenser, Walter Raleigh and Philip Sidney himself, and of the painter Nicholas Hilliard, made major discoveries in optics, algebra, biology and New World anthropology, largely because his friends and patrons demanded of him practical principles for the reconstruction of nature. We know, so the Elizabethans thought, by doing, and the best thing we can do is imitate the creative activity of nature.

So our assessment of the human resources that can be deployed for the work of restoration presents us with some curious but suggestive answers. First, we must look to the artistic and esthetic capacity, and learn more about its roots in our biology and evolution: the peculiar kinds of art and esthetics that are involved are those of the imitation of nature. Next, we may call on Renaissance esthetics for the nucleus of a sound body of theory about the artistic imitation of nature. Finally, we may find in the practice of art an example of the heuristic value of imitation. The attempt to reproduce accurately the functions of nature forces the artist not only to increasingly close observation, but beyond, to increasingly stringent experimental tests of ideas. This labor, so understood, is not merely analytical, but creative, and its natural reward is beauty.

### References

Callicott, J. B. (1984). The land esthetic. *Orion Nature Quarterly,* **3** (3), 16–23.
Marr, D. (1984). *Vision.* Van Nuys, California: Freeman.
Turner, F. (1986). *Natural Classicism: Essays on Literature and Science.* New York: Paragon House.

# II | Assembling whole systems in the field

Conference with them (first-rank artisans in the Venetian arsenal) has often helped me in the investigations of certain effects, including not only those which are striking, but also those which are recondite and almost incredible.

Galileo Galilei, *Two New Sciences*

Although rarely undertaken specifically for this purpose, restoration efforts have often proved to be of considerable value as a way of raising questions and testing ideas about the communities and ecosystems being restored. As a result, restoration efforts have frequently provided the background or opportunity for basic research reported in the literature, though this may not be acknowledged or, as in the case of the comments of Curtis & Partch quoted in Chapter 1, it may be acknowledged only glancingly. To some extent, this no doubt reflects the tradition of objectivity in the reporting of scientific results, which is characterized by an impersonal, expository style and the substitution of a section on "materials and methods" in place of any real account of how the research was actually carried out. (For further comments on this matter, see Peter Medawar's essay, *The art of the soluble* (London: Methuen, 1967, p. 151).)

Clearly an unfortunate result of this tradition is a danger that the discipline will lose sight of the origins of its own ideas. In particular, the contributions of adjacent disciplines, and ideas arising as a result of more immediately "practical" work may tend to be slighted. While in the absence of detailed historical analysis it is impossible to be sure about this, it seems quite plausible that as a result of something of this sort the contributions the practice of ecological restoration has made to basic ecological research have generally been underestimated.

In the chapters that follow, an attempt is made to explore this area and to consider what ecology may have derived from the experience of restoration, even when the restoration has not been carried out for explicitly experimental purposes. Represented here are communities of several types, and a general discussion by Anthony Bradshaw of the ecological value of restoring ecosystems that have been profoundly disturbed as a result of operations such as surface mining.

*A. D. Bradshaw*

Department of Botany, University of Liverpool

# 5 | The reclamation of derelict land and the ecology of ecosystems

Anyone who sets about to repair or restore any object, whether it is a clock, an old engine, or an eroded savanna, soon realizes that in order to do a good job he or she needs to know something about that object, the nature of its parts, how it was put together in the first place, and how the whole thing works. There is nothing subtle about this. Even the simplest repair demands some basic understanding, and the successful repair of a complex object may require a very detailed understanding (Fig. 5.1).

But repair or restoration really has two stages. The first is to discover and understand what is wrong. The second is then to put it right correctly and appropriately. How often are repairs of everyday objects unsuccessful because the repairer does not understand what is wrong or, having understood what is wrong, fails to make the repair correctly? As a result the repaired object either fails to work at all or fails again very quickly.

It is essential that we keep this analogy in mind when we approach any problem of land or ecosystem restoration. Like watch repair, restoration is a considerable intellectual challenge requiring that we understand not only the nature of the ecosystem itself, but also the nature of the damage and how to repair it. Hence my assertion in an earlier chapter that land restoration is an acid test of our ecological understanding. But there is a reverse side to this. What is discovered by experience to be critical for successful land restoration will also often be a critical contribution to ecological theory.

An examination of restoration practice should, therefore, have considerable heuristic and didactic value. Nowhere is this more apparent than in the reclamation of land that is so degraded that it can be termed derelict. Examples include mine wastes, surface mines where the soil has not been replaced, refuse disposal sites, urban clearance areas and old industrial sites. In many cases on such land the whole ecosystem has disappeared, not just

*Fig. 5.1.* Though often regarded as a purely technical challenge, the restoration of profoundly disturbed ecosystems provides unique opportunities for basic ecological research. Here, ecological restorationists study plantings of prairie species on a site covered by mine tailings in central Wisconsin (photograph by George F. Thompson, courtesy of Tom Hunt, Wisconsin Department of Natural Resources).

the biotic part but the soil as well. It is the ecological challenge of reclaiming such sites that will be considered in this chapter. However, before we can appreciate what this reclamation can contribute to our ecological understanding, we must realize that many professions can be involved. These represent a variety of approaches and points of view, some of which may be quite different approaches from those of ecologists.

### Starting points

To many people, especially civil engineers, the reclamation of derelict land is really nothing grandly scientific. It is a simple *technical* problem, a matter of finding permanent economical ways of achieving a few simple objectives: (1) stabilization of land surfaces, (2) pollution control, (3) visual improvement, and (4) general amenity, in order to preserve the structures in which they are interested and to prevent the land from being unpleasant to the people that use it. Also, since the land itself has value for what it can produce, we can add (5) productivity as a possible extra objective. Finally we can take on the more ambitious task of actually restoring the ecological communities that were present originally. This means adding (6) diversity, (7) species composition, and (8) ecosystem function.

In any particular situation only one or two of these objectives may have to be met. Highway engineers, for instance, are looking only for stabilization and visual improvement, but Australian mineral sand miners are looking not only for these but also for everything else, because they are required to put back exactly what was there previously and to ensure that it functions properly. All of these examples constitute land reclamation, but only the latter can be considered as full restoration.

There is a considerable difference, however, between recognizing an objective and understanding how to achieve it. To achieve any or all of these objectives, we are presuming that vegetation will be employed (although it must be remembered that there may be other solutions, such as a layer of concrete). And if vegetation is to be developed, then the matter is essentially an *ecological* problem – that of the reconstruction of ecosystems – because that is what will exist if stable, self-sustaining vegetation is achieved.

This vegetation will have two basic qualities: first its structure, and second its functioning. The structure, measured in terms of species diversity and physical and biological complexity, can be simple – a monospecific salt marsh grass sward, for example. Or it can be as complex as a tropical rainforest or a South African fynbos heathland. Similarly, the functioning or ecological "metabolism" of the system, usually measured in terms of processes such as productivity and nutrient cycling, can be low or high. In any case, to put all this back together in working order is a considerable challenge.

Where land has been destroyed by mining or a similar activity, what is

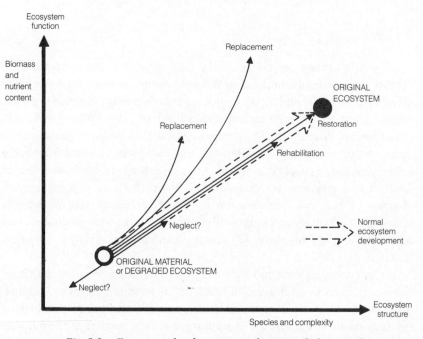

*Fig. 5.2.* Ecosystem development can be quantified in two dimensions, structure and function. In natural succession there is an increase in both dimensions. When ecosystems are degraded by mining or other operations, there is reduction in both dimensions, perhaps almost to zero. The first option with such a degraded or derelict ecosystem is to do nothing, in which case it may recover slowly by natural processes, though it may also degrade further by erosion or landslip. The second option is to try to build back exactly what was there before. If this is successful then what is achieved is "restoration". If it is not completely successful then what is achieved can be termed "rehabilitation." The third option is "replacement", in which an alternative to the original ecosystem is produced. This could be simpler in structure but more productive (replacement of a woodland by an agricultural grassland, for example). Or it could be simpler and less productive (if, for example, the woodland were replaced by an amenity grassland). The crucial point is that any development of an ecosystem on degraded or derelict land will require carefully planned assistance if the development is to occur quickly (Bradshaw, 1984).

left is something exceedingly basic – just the raw materials, the skeleton from which a soil can be formed, and which, with plants and animals, will develop into an ecosystem. Ecologists will recognize that this process of development, whether it is allowed to occur naturally or is artificially assisted, is the process of "primary succession". We can represent what is involved ecologically in land restoration by a simple, two-dimensional graph (Fig. 5.2) originally proposed by Magnuson *et al.* (1980) and developed subsequently (Bradshaw, 1984). Despite the obvious simplification, this graph has merit because it provides a much needed ecological perspective on restoration practice, and also because it distinguishes the different forms that reclamation of degraded land may take.

In some situations the degradation of the ecosystem may not have been complete, and the soil may have been left partially intact. In this case, we would be dealing with the process of *secondary succession*. On the graph the starting point would be further from the origin, so there would be less distance to progress to reach the desired endpoint.

But we must also consider land reclamation from a practical point of view. In almost all operations, whatever the endpoint, the following are important considerations: (1) speed of attainment, (2) cheapness, (3) reliability in attainment and (4) stability (implying no maintenance or minimum maintenance of the final product). Except for speed, nature meets these criteria unassisted through natural succession. In land reclamation it is therefore important for us to understand, and as far as possible to take advantage of, what occurs in natural successions, since our aims are generally modeled on the achievements of nature. Equally, however, the reciprocal applies. In ecology it is important for us to understand and to incorporate into our theoretical ecological framework what is found to be successful and important in land restoration, since what is found to be important must be a crucial determinant in ecosystem functioning. In this way ecosystem reconstruction should be a meeting place between practitioners and theorists.

Yet, as in agriculture, until recently there has been a great divorce between these two approaches. To be blunt, ecologists have thought of land reclamation as beneath their dignity, and land reclaimers have not seen any value in an ecological approach. This is unfortunate because, on the one hand, ecologists have missed opportunites to enlarge and test their theories, and on the other hand, land reclaimers could have been saved unnecessary costs and even failures.

This duality and reciprocity becomes clear when the actual steps of the reconstruction process are considered in more detail.

## Problems in the reconstruction process

The starting point must be the soil, or at least the substrate into which plants must establish and root, for although soil can exist without plants, there are few plants that can exist without soil. This starting material is usually extremely skeletal on derelict sites, and its properties and situation determine in a very crucial manner the degree to which an ecosystem can develop naturally on the site, how far this development will progress, and what treatments are necessary to assist its development.

Although there are vast numbers of starting materials, ranging from rock faces left by limestone quarrying to red mud produced by aluminium refining, occurring in situations from tundra to tropical forest, the problems they pose can always be reduced to four (Table 5.1). The universality of this table is important in view of the fact that reclaimers often insist that their own problem is unique. It reflects the very simple needs of plants for: (1) a medium into which they are physically able to root, (2) an adequate water supply, (3) an adequate nutrient supply, and (4) lack of toxicity.

One of the simplest ways by which it is possible to see the importance of these factors is by examining the natural colonization that occurs on the poorest of materials. Both colonization and ecosystem development are always slow on severely degraded substrates. However, they are not usually any slower than the same processes occurring on similar materials of natural origin, such as on the glacial moraines so well described by Crocker & Major (1955), where woodland appears in 30–70 years. Similar speeds of development have been found on ironstone wastes by Leisman (1957), and on kaolin wastes by Roberts *et al.* (1981).

However, in several instances ecosystem development can be much slower. Woodland may not appear even after 100 years on some colliery spoils in England (Hall, 1957), and also on lime wastes from the Leblanc process (Bradshaw, 1983). In these cases the slowness of the succession can be attributed to specific arresting factors: very low pH on the colliery spoil and extreme phosphorus deficiency due to high pH and high calcium concentration on the lime wastes. In quarries, Park (1982) has shown that certain combinations of drought and surface texture can result in dramatic reductions in colonization rates. Such situations are completely analogous to what can occur naturally. For instance, on the raw materials of the new volcanic island of Surtsey, large parts still have no higher plants ten years after the island's formation. Part of this is due to difficulties of immigration, but much of it is due to the extreme nature of the volcanic substrates (Fridriksson, 1975).

Table 5.1. *The underlying problems of derelict land and their treatment (from Bradshaw, 1983)*

| Category | Problem | Immediate treatment | Long-term treatment |
|---|---|---|---|
| **Physical** | | | |
| Structure | Too compact | Rip or scarify | Vegetation |
| | Too open | Compact or cover with fine material | Vegetation |
| Stability | Unstable | Stabilizer/mulch | Regrade or vegetation |
| Moisture | Too wet | Drain | Drain |
| | Too dry | Organic mulch | Vegetation |
| **Nutrition** | | | |
| Macronutrients | Nitrogen | Fertilizer | Legume |
| | Others | Fertilizer + lime | Fertilizer + lime |
| Micronutrients | | Fertilizer | |
| **Toxicity** | | | |
| pH | Too high | Pyritic waste or organic matter | Weathering |
| | Too low | Lime or leaching | Lime or weathering |
| Heavy metals | Too high | Organic mulch or metal-tolerant cultivars | Inert covering or metal-tolerant cultivars |
| Salinity | Too high | Weathering or irrigation | Tolerant species or cultivars |
| **Plants and animals** | | | |
| Wild plants | Absent or slow colonization | Collect seed and sow or spread soil containing propagules or plants | Ensure appropriate conditions |
| Cultivated plants | Absent | Sow normally or hydroseed | Appropriate aftercare |
| Animals | Slow colonization | Introduce | Ensure appropriate habitat |

Where specific limiting (arresting) factors are not obvious, however, as on glacial moraines or on the inert sand waste produced by the kaolin industry in south-western England, succession still usually occurs at a slow rate. This raises numerous questions about what the rate-limiting factors are and how they vary from site to site. If we knew the answers we would perhaps be in a better position to achieve efficient restoration.

Investigations suggest that nitrogen is usually the major limiting factor (Crocker & Major, 1955; Roberts *et al.*, 1981). It is the only nutrient that changes significantly with ecosystem development and that can be shown to be continually limiting (Marrs *et al.*, 1983, and Ch. 16).

But other factors can be equally limiting, or may at least play a part in limiting the rate of ecosystem development. They can be physical, chemical, or biological. One of the most interesting is the biological factor of immigration. It is difficult to believe that the supply of suitable propagules could be important in determining ecosystem development on a patch of degraded land surrounded by a normal varied agricultural countryside. But where the substrate of that degraded land is alien and very different from the natural soils of the immediate region, it is possible that the ecologically appropriate species are just not present in the vicinity. In this situation the only species that will be able to colonize will be those with special powers of long-range dispersal. In other words, these sites have the properties of islands and show similar biogeographic characteristics (Gray, 1982), including a direct relationship between the area of the site and the number of species that come to occupy it. What is crucial is that the whole process of ecosystem development can be held up by the lack of suitable colonists (Bradshaw, 1983), again just as in nature, on new islands such as Surtsey, for example.

This can be tested directly. In recent experiments, a large number of missing species has been deliberately introduced onto alkali wastes in north-western England. Some of these species, notably *Blackstonia perfoliata* and *Rhinanthus minor*, have spread rapidly over the sites, clearly showing that ecosystem development is being held up by lack of colonists. It is interesting to find that other missing species, such as *Primula veris*, either failed to establish or would grow only in the presence of added major plant nutrients, indicating that chemical limiting factors also play an important role (Ash & Bradshaw, unpublished data).

**Treatments**

Progress in land restoration technology during the past 20 years has been remarkable. There are now very few degraded environments in which the original ecosystem, or an effective substitute, cannot be established. Unfortunately, much of the detailed methodology is available only in special publications, though there are two substantial reviews (Schaller & Sutton, 1978; Bradshaw & Chadwick, 1980). Nevertheless we must not run away with the idea that all such restoration efforts will always be perfect (in the narrow sense). The products may be deficient in either structure or function. And while a study of the way success is achieved is very instructive, encountering difficulties and attempting to account for them may be even more so.

*Soil replacement*

The simplest treatment for any existing area of degraded land is to disregard its individual problems and import a new soil surface, on which a new ecosystem can quickly be established. In progressive mining operations, it is of course becoming mandatory for surface soils to be conserved and replaced. In the restoration of small, existing degraded areas, the use of a layer of topsoil to cover up whatever is wrong beneath is common practice.

In a crude sense, this may require little understanding of what is wrong with the site, but it turns out that it does raise many interesting questions about soils and about the relationships between soils and plants. Thus we have come to realize that the processes of importing and spreading, or of conservation and replacement, do demand considerable care to ensure that the topsoil and subsoil structures, especially their macrostructure, are not damaged. In most cases the critical factors involved here are drainage and retention of water (Jansen, 1981). But in extreme cases, soil crumb structure, normally built up over a long period by natural soil processes, can be so damaged in the restoration process that rooting is restricted. Although more work is needed in this area, there is already sufficient evidence to make it clear that the relationship between soil structure and rooting depth is of considerable ecological significance. As a result, quite elaborate systems of soil replacement are now being investigated (Department of Environment, 1982), and quite precise recommendations are suggested to obtain maximum productivity in agricultural ecosystems (Coppin & Bradshaw, 1982). What is necessary for other, more "natural" ecosystems has not yet been precisely defined, but all the evidence so far suggests that the soil part of any

62     *A. D. Bradshaw*

ecosystem can have important physical properties that ought to be considered more by ecologists (see Ch. 7).

But this is not all, since it is commonly found that the performances (measured as yield) of ecosystems reconstructed on topsoil are inadequate. Experimentation shows that the soil used is unable to sustain vigorous growth unless suppled with extra nitrogen (Bloomfield, Handley & Bradshaw, 1981). This can be due either to deterioration of the soil during storage between gathering and respreading, or to excessive amounts of subsoil being included in material sold as "topsoil". Either way, the fragility of soil fertility, especially in relation to its ability to provide nitrogen, is made very clear by reclamation experience, a point we shall return to later. (Other ecological phenomena that have been clarified as a result of experience with soil handling procedures are discussed in Ch. 14.)

### Direct treatment

In many situations, however, soil cannot be imported or replaced, and the material existing on a site has itself to be treated directly to achieve restoration. The range of treatments commonly used is summarized in Table 5.1. These treatments can be very effective in allowing rapid ecosystem development. This is itself very instructive, not just because it shows that ecosystem restoration is perfectly feasible but because it illustrates some of the distinct requirements for ecosystem development. If any of these requirements is not met because a particular treatment is omitted or because it is inadequate, ecosystem development can fail or be delayed. This may seem rather obvious, but the ways in which such failures occur can provide insights into the requirements of ecosystems and their functioning that may otherwise be very difficult to obtain.

### Physical treatments

The need to overcome physical factors in establishing vegetation is taken for granted by all practitioners. Considerable efforts are made to loosen subsoil and topsoil by ripping the substrate to depths up to 1 meter. This is then followed by shallower cultivations, to about 20 cm. Finally, the soil surface is carefully prepared to provide a good seedbed, a favorable season is chosen for sowing, and the seed is buried by mechanical means. Such artificial treatments in themselves suggest the critical nature of the seed germination and seedling establishment process. Sometimes these basically agricultural treatments are difficult to apply, however, and a different and in many ways more natural system of sowing is used – the hydraulic seedling technique, in which the seed is spread in a slurry containing water, fertilizer, mulch, and stabilizer. This method is more natural in the sense that the seed is just scattered on the soil surface and not buried. Often, however, the slurry includes extra, unnatural components – the fertilizer, mulch, and stabilizer – which are intended to help seedling establishment. Interestingly, a detailed analysis shows that only the mulch has positive effects. The other components can be profoundly deleterious, especially in their effect on legumes, and when the climate is not optimal (Roberts & Bradshaw, 1985). It is also interesting that in many cases a rough soil surface is better than any other treatment. This fits in with studies on seed-soil surface relationships (Harper & Benton, 1966), and also suggests ways of economizing in seedbed preparation, since a rough soil surface is much cheaper to provide than a finely prepared seedbed.

Subsequent growth of the ecosystem might appear to be rather unrelated to physical factors in the substrate. However, in the reclamation of raw colliery spoil heaps in England, which contain a high clay fraction, it has recently been shown that soil texture, through its effect on structure, can continue to reduce grass yields for many years (Rimmer, 1982; Elias *et al.*, 1982). What happens is that, following careful preparation by ripping and cultivation, the soil collapses back to become a very hard and dense medium through which plant roots cannot easily penetrate. There are, however, natural processes which obviate this, the best example of which comes from the changes observed by Crocker & Major (1955) on glacial moraines in Alaska, where soil has slowly become open and more porous. This is due to incorporation of organic matter and to the activities of soil flora and fauna. It seems that we are only now beginning to appreciate the importance of these changes and have not yet discovered how to harness them for practical use. Yet there is one remarkable experiment on the effects of earthworms on the characteristics of the soils of newly reclaimed polders in the Netherlands that provides clear evidence of what can be achieved. Where the earthworms have been introduced, the surface mat has disappeared and infiltration capacity has increased more than a hundredfold (Hoogerkamp, Rogaar & Eijsackers, 1983; see also Ch. 7).

### Nutrient addition

Having dealt with, or at least worried about, physical problems, the practitioner next expects to have to deal with a series of problems connected with nutrient supply. The need for nutrient additions when ecosystems are being reconstructed on skeletal material is obvious. Materials such as mine wastes or subsoils are unlikely to contain any significant quantities of nitrogen in particular, since in most ecosystems nitrogen accumulates in surface soils as a result of biological activity. But other important nutrients such as phosphorus may also be deficient, either in terms of total or available amounts.

It is easy to provide the necessary nutrients by means of ordinary agricultural fertilizers containing nitrogen, phosphorus, and potassium. In fact, the effects of such supplements are nearly always dramatic, making it very clear that ecosystems, even in a juvenile condition, do need considerable amounts of nutrients for growth, and that these substances are likely to be limiting in most of the skeletal materials found on severely degraded land. Moreover, practical experience, including occasional disastrous failures, shows that these nutrients must be provided repeatedly or growth is substantially reduced. This raises an interesting question: why should the nutrient requirement be so persistent?

If a simple factorial aftercare experiment is carried out, in which all combinations of the three major nutrients (nitrogen, phosphorus and potassium) are applied to a newly established ecosystem that has become moribund, nitrogen is almost always found to be deficient. If nitrogen is provided, growth continues, but if in a subsequent year the treatment is not repeated, growth stops (Bloomfield, Handley & Bradshaw, 1982) (Fig. 5.3). The reasons for this continuing need for nitrogen can be discovered by examining the nutrient requirements of ecosystems. These can easily be calculated in relation to various levels of productivity (Table 5.2). An adequate productivity for many ecosystems is about 5000 kg ha$^{-1}$ yr$^{-1}$, which requires approximately 100 kgN ha$^{-1}$ – more than any other nutrient. But what is unique about nitrogen is not only that it does not occur as a soil mineral, but also that it is supplied by the decomposition of the organic matter contained in the soil. The amount of the mineral nitrogen that would be supplied by different rates of decomposition and different soil capitals can be calculated (Table 5.3). In temperate regions, a typical annual decomposition rate is about 1/16.

From this it follows that to maintain a productivity of about 5000 kg ha$^{-1}$ yr$^{-1}$ in temperate regions, a soil nitrogen capital of about 1600 kg ha$^{-1}$ is required. The exact amount must depend on rates of decomposition and the particular type of vegetation being considered, but this figure is confirmed by what is found in natural successions, although in some situations, such as on kaolin waste, the value may be somewhat less (Marrs & Bradshaw, 1982). In land reclamation the annual fertilizer dressings

Table 5.2. *Annual nutrient requirement for ecosystems with different productivities (Bradshaw, 1983)*

| | Nutrient content assumed (per cent) | Production level (kg ha$^{-1}$ year$^{-1}$) | | | |
|---|---|---|---|---|---|
| | | 1000 | 5000 | 10 000 | 20 000 |
| Nitrogen | 2.0 | 20 | 100 | 200 | 400 |
| Potassium | 1.1 | 11 | 55 | 110 | 220 |
| Magnesium | 0.51 | 5.1 | 26 | 51 | 102 |
| Calcium | 0.26 | 2.6 | 13 | 26 | 52 |
| Phosphorus | 0.18 | 1.8 | 9.0 | 18 | 36 |
| Type of ecosystem | | Tundra and desert | Poorly productive temperate | Productive temperate and poorly productive tropical | Tropical |

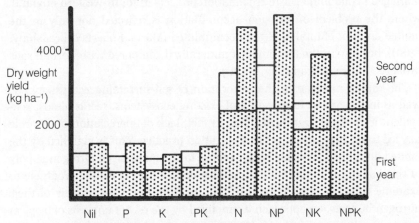

Fig. 5.3. When almost any ecosystem has been established on very degraded soils with the assistance of fertilizer treatments, growth declines dramatically if these treatments are terminated after one or two years, which is a common practice. The reason for this is apparent from this experiment, in which factorial combinations of nitrogen, phosphorus, and potassium have been added to a moribund grass sward established during the reclamation of colliery spoil about four years earlier. There is clearly a gross deficiency of nitrogen, rather than anything else, and this persists even in the second year of treatment. The nitrogen supply in an ecosystem is derived from a capital of organic nitrogen in soil. This experiment, and others, indicates that the size of that capital is crucial in developing ecosystems and that in restoration practice it is essential to devise ways in which it can be built up rapidly (Bloomfield, Handley & Bradshaw, 1982). (Shaded: growth with fertilizer; white: growth without.)

Table 5.3. *The organic soil nitrogen capital needed ($kg\ N\ ha^{-1}$) to satisfy different nitrogen requirements, assuming various decomposition rates (Bradshaw, 1983)*

| | Decomposition rate | | | |
|---|---|---|---|---|
| Annual requirement | 1/64 | 1/16 | 1/4 | 1 |
| 200 | 12800 | 3200 | 800 | 200 |
| 100 | 6400 | 1600 | 400 | 100 |
| 50 | 3200 | 800 | 200 | 50 |
| Type of ecosystem | Montane | Cool temperate | Warm temperate | Tropical |

provided do not contain much more than 100 kg N/ha$^{-1}$, some of which is lost by leaching. So it is clear that the necessary soil capital will take some time to accumulate. This is confirmed by what is found even in well-managed kaolin mine waste reclamation projects in south-western England, where the problem of nitrogen accumulation is reflected not only in the limited amount of total nitrogen accumulated (Marrs, Roberts & Bradshaw, 1980), but also in the low amount mineralized compared with normal soils (Roberts, Marrs & Bradshaw, 1980).

The role of nitrogen in the restoration of self-sustaining ecosystems, as well as in the proper functioning of existing ecosystems, seems almost self-evident from these arguments. But, in fact, lack of appreciation of this role has led to numerous failures in restoration practice. This is matched by the limited discussion that has been devoted to the question of nitrogen supply in relation to the functioning of natural ecosystems. It will be profitable to examine natural ecosystems in the future from the point of view of their nitrogen budgets, an approach exemplified by the recent analysis of tropical montane rainforest by Grubb (1977a), of prairie by Woodmansee (1979), and especially of temperate forests where productivity may be closely related to soil mineral nitrogen supply (Nadelhoffer, Aber & Melilo, 1983). (For further discussion of this point, see Ch. 16.)

In ecosystem restoration there is obviously a need to find ways other than repeated fertilizer applications to provide the nitrogen capital required. An obvious approach is to use legumes and other nitrogen-fixing species such as alder, because these can readily fix more than 100 kg N ha$^{-1}$ yr$^{-1}$ even under difficult conditions. Another, increasingly important technique is the application of various nitrogen-rich wastes such as sewage sludge or the fine fraction from pulverized domestic refuse. In general, the role of nitrogen–

fixing species in primary successions has long been recognized. But it is perhaps the evidence of their crucial role in restoration that is refocusing our attention on their importance in natural situations.

The other nutrient that is very likely to be deficient is phosphorus. Some substrates, such as lime wastes and colliery spoils, are not only initially deficient but have high phosphorus sorption capacities (Fitter, 1974). This not only explains why natural ecosystem development can be so slow on these wastes (Bradshaw, 1983), but has also led to raising phosphorus applications to as much as 200 kg/ha$^{-1}$ in the restoration of some materials. Again, though agriculturists have long recognized the crucial role of phosphorus in limiting yield, it is surprising how rarely ecologists have investigated its role in natural systems.

## Treating toxicities

The drastic ecological effects of toxicity on degraded or derelict land are demonstrated very clearly by areas contaminated by heavy metals. These can remain nearly bare, colonized only by a sparse vegetation of metal-tolerant plants for periods of time that may be measured in centuries (Ernst, 1974; Bradshaw & McNeilly, 1981). Clearly, attempts to re-establish vegetation in these areas without dealing with the problem of metal toxicity are doomed to failure. (The evolution that has taken place on these sites is itself a fascinating story, summarized by McNeilly in Ch. 18.)

Experiments on these wastes show that plant growth depends on a combination of metal tolerance in the populations of the colonizing species and the presence of sufficient nutrients, especially nitrogen and phosphorus. The ecologically elegant solution for reclamation is therefore to use metal-tolerant plants and to supply fertilizer. This can be very successful (Smith & Bradshaw, 1979). It truly copies nature, but, like nature, it also results in an odd and very simplified ecosystem in which nutrient cycling is so much reduced by the heavy metal toxicity that growth is limited unless repeated fertilizer applications are given (McNeilly, Williams & Christian, 1984). Such sites clearly demonstrate the importance of properly functioning nutrient-cycling systems.

To obtain a normal ecosystem the effects of the metals have to be dealt with completely. A covering of organic matter to bind any available metals was originally thought to be effective. But this effect declines after a few years as the organic matter disappears and the toxicity returns. As a result, the main emphasis has turned to covering toxic substrate layers with an inert layer to serve as a barrier to the upward movement of metals and the downward growth of roots (Williamson, Johnson & Bradshaw, 1982; Smith, 1985).

In the case of metal-contaminated sites, therefore, the nature of the toxicity is such that direct treatment is not completely satisfactory. Once again what is required for successful restoration is indicated by the vegetation colonizing natural, undisturbed metal-contaminated areas. These can be recognized by their anomalous vegetation (Nicholls *et al.*, 1965), even when this vegetation has had many thousands of years in which to develop.

### Adding species

In natural ecosystem development, species invade slowly and can take advantage of the developing environment produced by physical and chemical changes that occur during primary succession. They can also take advantage, so to speak, of years when conditions for colonization are especially favorable. Both of these effects are apparent in the natural colonization of kaolin wastes. Nitrogen-accumulating legumes, such as gorse (*Ulex europaeus*), which represent an essential stage in the development of woody species on such sites, themselves colonize only in years when there is a wet spring (Dancer, Handley & Bradshaw, 1977).

In general, the establishment of species by natural processes tends to be slow and stochastic. It also depends on the source of propagules available in the vicinity. So in artificial restoration the required species are introduced artificially and sown by ordinary agricultural or forestry techniques. The choice of species can be tailored to suit the ecosystem being reconstructed, including species suitable for early as well as late stages of ecosystem development; also nitrogen fixers and other appropriate plants, trees, shrubs, and herbs. For practical reasons all the species are usually sown or planted at the same time. This may sound rather simple, but in practice to compress the normal, gradual process of establishment into a single period may actually be difficult.

An important consideration here is the provision of microenvironments for establishment which are suitable, both chemically and physically, for the desired species. There are many ways this might be done. The most common ways are to provide one or more nurse species, and appropriate fertilization. But it can be difficult to know what is appropriate for one particular species, let alone for a whole group of different ones. Some species of *Calluna* and *Erica*, for example, are moorland species normally found in open, very nutrient-deficient habitats. But on kaolin sand wastes, it is found, quite remarkably, that the combination of both a nurse grass and fertilizer leads to the best re-establishment of these two species from dormant seed in a thin scattering of topsoil (Fig. 5.4). The provision of an extra subsoil layer does not have any extra effect. The nurse grass must somehow be providing a suitable microenvironment. The effect of the fertilizer suggests that the early seedling stages are more sensitive to nutritional deficiencies than later stages. Despite its possible competitive effects, the provision of a nurse species in combination with nutrient addition is found to be important for establishment in many situations where ecosystems are being established. This is true even when the aim is to restore faithfully a pre-existing ecosystem, such as coastal heath after mineral sand mining in Australia (Brooks, 1976). In other areas (the Piceance Basin, for example) irrigation may also be valuable (Doerr, Redente & Sievers, 1983). (The importance of

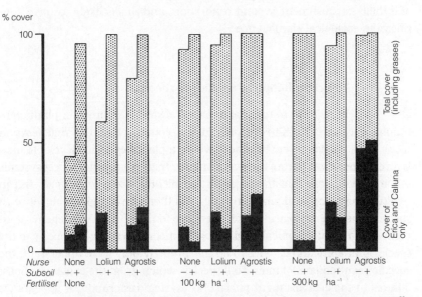

*Fig. 5.4.* The establishment phase in land restoration is especially important when wild species are being replaced. In this experiment two heathers (*Calluna vulgaris* and *Erica cinerea*) are being established on very sandy kaolin wastes in south-western England by collecting litter from natural heather moorland and spreading it on the waste surface. The effects of a nurse grass (*Agrostis capillaris*), fertilizer, and a thin covering of subsoil are being tested. Despite the fact that the heathers naturally grow in very exposed and nutrient deficient soils, establishment is best after three years on plots treated with a combination of high fertilizer and the nurse grass. Such results not only provide information useful in refining restoration techniques, but also help define more accurately the regeneration niche of the species involved (original data of P. D. Putwain & A. D. Gilham).

appropriate micro-environments for seedling establishment is discussed further by Gross in a subsequent chapter.)

All this suggests that the hard experience of land restoration is beginning to provide good evidence for the interdependence of species in succession. Sequential planting, which is now being found to be useful, will provide further evidence. If tree lupins, for instance, are sown in advance of forest tree planting on degraded sand dunes in New Zealand, tree growth is substantially improved (Gadgil, 1971). The trees are introduced only after the lupins have reached the mature stage, when their nitrogen accumulation has reached a maximum. On kaolin wastes, it is found that if annual grasses and legumes are sown a year ahead of perennial grasses and legumes, the final establishment of the latter is much better. There is still a great deal more practical experience to be gained, however, about species and their establishment in land restoration, and this is likely to have considerable ecological significance.

### The contribution of practice to theory

In relation to the maintenance of species richness in plant communities, Grubb (1977b) introduced the concept that the regeneration niche – the particular environment where a plant begins its development – is a very important part of the overall niche. In the restoration of ecosystems one might suppose that the regeneration niche space is wide open. But in reality it may lack particular attributes and therefore be unsuitable for most species. The first ecological step is therefore to manipulate the regeneration niche by physical, chemical, and biological means, to tailor it properly to the species that are wanted. We can do this properly only if we understand the specific requirements of individual species, which, from the evidence of past failures in the establishment phase of ecosystem restoration, it appears we have not always managed to do. These failures can be expensive and may be discouraging, but they do provide opportunities for defining exactly what we do not yet know about the requirements of species and ecosystems. At the same time the general accumulated experience of restoration should make major contributions to general theories of community development.

Succession, for example, is an important ecological concept. Yet the mechanics of succession under various conditions are still widely debated. There are good recent discussions by Connell & Slatyer (1977) and Miles (1979). For our purposes it is best to be very simple. Perhaps the first consideration is our understanding of the origin of the environmental changes that lead to the sort of natural recovery shown in Figure 5.1. This leads to improvement in the environment. Perhaps almost the first point that ecologists discuss is whether these changes are allogenic or autogenic in origin, that is, whether they arise from outside the community (for example, weathering, leaching) or from within it (from plant roots, legumes, etc.). It is clear that in ecosystem restoration both types of factors play important roles.

But the aim of the restorationist is to accelerate succession, and it is clear that these factors are rarely enough by themselves to allow rapid ecosystem development and that considerable artificial help must usually be given. To provide this help, however, it is necessary to understand the factors limiting succession at each point of its progress and to relieve them by cultivation, fertilization, liming, or other specific treatment. There is a need to optimize the environment, both for individual species and for the entire ecosystem. This is because both species and ecosystems have specific physical and chemical requirements, which in natural ecosystem development may be satisfied only after a lapse of time (see Ch. 7).

As a result, ecosystem restoration provides a way of seeing more clearly the needs of ecosystems and the importance of both allogenic and autogenic factors in primary succession. In particular, the contribution of autogenic factors which can be demonstrated in the course of restoration work indicates the importance of *facilitation*, in the sense of Slatyer (1977). This is the idea that one species makes it easier for a second species to participate in a particular succession, an assertion that has at times been questioned (Drury & Nisbet, 1973).

Facilitation leads to "relay floristics", the concept that there is a specific sequence of species (Egler, 1954). One of the most interesting aspects of ecosystem restoration, the significance of which has hardly been explored, is the practical need to compress establishment into one phase so that relay floristics cannot occur. For this reason the results of techniques such as the establishment of trees by seed as part of the initial treatment (Luke, Harvey & Humphries, 1982) will be of great interest. As might be expected, the results have been rather variable so far.

In this discussion little attention has been paid to the influence of competition in restoration practice yet it certainly must occur. Indeed there is evidence that it does occur, an example being the deleterious effects of grass on tree establishment (Bradshaw & Chadwick, 1980). But in general we have not paid enough attention to competition as a critical factor, and it will be discussed in later chapters by Gilpin, Gross, Rosenzweig, and Kline & Howell. If we include predation as part of competition, then we must report almost total ignorance, which, in view of the crucial role of such factors in natural ecosystems, is certainly unsatisfactory.

In conclusion, it is clear that natural succession has many probabilistic components, and that in ecosystem restoration these elements have been, or should be, removed as far as possible. At the same time the whole construction process has to be hastened. This requires us to understand ecosystem needs and functions very precisely and in practical terms. Ecological theory has much to contribute to this practical goal, but at the same time restoration practice has much to contribute to ecological theory by providing not only experimental evidence but also an acid test of our understanding.

Ecologists working in the field of ecosystem restoration are in the construction business and, like their engineering colleagues, can soon discover if their theory is correct by whether the airplane falls out of the sky, the bridge collapses, or the ecosystem fails to flourish.

### References

Bloomfield, H. E., Handley, J. F. & Bradshaw, A. D. (1981). Topsoil quality. *Landscape Design*, **135**, 32–4.

Bloomfield, H. E., Handley, J. F. & Bradshaw, A. D. (1982). Nutrient deficiencies and the aftercare of reclaimed derelict land. *Journal of Applied Ecology*, **19**, 151–8.

Bradshaw, A. D. (1983). The reconstruction of ecosystems. *Journal of Applied Ecology*, **20**, 1–17.

Bradshaw, A. D. (1984). Ecological principles and land reclamation practice. *Landscape Planning*, **11**, 35–48.

Bradshaw, A. D. & Chadwick, M. J. (1980). *The Restoration of Land*. Oxford: Blackwell.

Bradshaw, A. D. & McNeilly, T. (1981). *Evolution and Pollution*. London: Arnold.

Brooks, D. R. (1976). Rehabilitation following mineral sand mining on North Stradbrooke Island, Queensland. In *Landscaping and Land Use Planning as Related to Mining Operations*, pp. 93–104. Adelaide: Australasian Institute of Mining and Metallurgy.

Connell, J. H. & Slatyer, R. D. (1977). Mechanisms of succession in natural communities and their role in community stability and organisation. *American Naturalist*, **111**, 1119–44.

Coppin, N. J. & Bradshaw, A. D. (1982). *Quarry Reclamation*. London: Mining Journal Books.

Crocker, R. L. & Major, J. (1955). Soil development in relation to vegetation and surface age at Glacier Bay, Alaska. *Journal of Ecology*, **43**, 427–48.

Dancer, W. S., Handley, J. F. & Bradshaw, A. D. (1977). Nitrogen accumulation in kaolin mining wastes in Cornwall. I. Natural communities. *Plant and Soil*, **48**, 153–67.

Department of Environment (1982). *Bush Farm Working Party, Joint Agricultural Land Experiment Program Report No 2 for Bush Farm, Upminster, Essex*.

Doerr, T. B., Redente, E. F. & Sievers, T. E. (1983). Effect of cultural practices on seeded plant communities on intensively disturbed soils. *Journal of Range Management*, **36**, 423–8.

Drury, W. H. & Nisbet, I. T. (1973). Succession. *Journal of the Arnold Arboretum*, **54**, 331–68.

Egler, F. E. (1954). Vegetation science concepts. I. Initial floristics composition, a factor in old-field development. *Vegetatio*, **4**, 412–17.

Elias, C. O., Morgan, A. L., Palmer, J. P. & Chadwick, M. J. (1982). *The Establishment, Maintenance and Management of Vegetation on Colliery Spoil Sites.* University of York: Derelict Land Research Unit.

Ernst, W. (1974). *Schwermetallvegetation der Erde.* Stuttgart: Fischer.

Fitter, A. H. (1974). A relationship between phosphorus requirements, the immobilisation of added phosphate, and the phosphate buffering capacity of colliery shales. *Journal of Soil Science*, **25**, 41–50.

Fridriksson, S. (1975). *Surtsey*. London: Butterworth.

Gadgil, R. L. (1971). The nutritional role of *Lupinus arboreus* in coastal sand dune forestry. I. The potential influence of undamaged lupin plants on nitrogen uptake by *Pinus radiata*. *Plant Soil*, **34**, 357–67.

Gray, H. (1982). Plant dispersal and colonisation. In *Ecology of Quarries*, ed. B. N. K. Davis, pp. 27–31. Cambridge: Institute of Terrestrial Ecology.

Grubb, P. J. (1977a). Control of forest growth and distribution on wet tropical mountains. *Annual Review of Ecology and Systematics*, **8**, 83–107.

Grubb, P. J. (1977b). The maintenance of species richness in plant communities and the importance of the regeneration niche. *Biological Reviews*, **52**, 107–45.

Hall, I. G. (1957). The ecology of disused pit heaps in England. *Journal of Ecology*, **45**, 689–720.

Harper, J. L. & Benton, R. A. (1966). The behaviour of seeds in soil, part 2. The germination of seeds on the surface of a water supplying substrate. *Journal of Ecology*, **54**, 151–66.

Hoogerkamp, M., Rogaar, H. & Eijsackers, H. F. (1983). The effect of earthworms (Lumbricidae) on grassland on recently reclaimed polder sods in the Netherlands. In *Earthworm Ecology*, ed. J. E. Satchell, pp. 85–104. London: Chapman & Hall.

Jansen, I. J. (1981). Reconstructing soil after surface mining of prime agricultural land. *Mining Engineering*, March 1981, 312–4.

Leisman, G. A. (1957). A vegetation and soil chronosequence on the Mesabi Iron Range spoil banks, Minnesota. *Ecological Monographs*, **27**, 221–45.

Luke, A. G. R., Harvey, H. & Humphries, R. N. (1982). The creation of woody landscapes on roadsides by seeding – a comparison of past approaches in West Germany and the United Kingdom. *Reclamation and Revegetation Research*, **1**, 243–53.

Magnuson, J. J., Regier, H. A., Christien, W. J. & Sonzogi, W. C. (1980). To rehabilitate and restore Great Lakes ecosystems. In *The Recovery Process in Damaged Ecosystems*, ed. J. Cairns, pp. 95–112. Michigan: Ann Arbor Science.

Marrs, R. H. & Bradshaw, A. D. (1982). Nitrogen accumulation, cycling and the reclamation of china clay wastes. *Journal of Environmental Management*, **15**, 139–57.

Marrs, R. H., Roberts, R. D. & Bradshaw, A. D. (1980). Ecosystem development on reclaimed china clay wastes. I. Assessment of vegetation and capture of nutrients. *Journal of Applied Ecology*, **17**, 709–18.

Marrs, R. H., Roberts, R. D., Skeffington, R. A. & Bradshaw, A. D. (1983). Nitrogen and the development of ecosystems. In *Nitrogen as an Ecological Factor*, ed. J. A. Lee, S. McNeill & I. H. Rorinson, pp. 113–36. Oxford: Blackwell.

McNeilly, T., Williams, S. T. & Christian, P. J. (1984). Lead and zinc in a contaminated pasture at Minera, N. Wales and their impact on productivity and organic matter breakdown. *The Science of the Total Environment*, **38**, 183–98.

Miles, J. (1979). *Vegetation Dynamics*. London: Chapman & Hall.

Nadelhoffer, K. J., Aber, J. D. & Melilo, J. M. (1983). Leaf–litter production and soil organic matter dynamics along a nitrogen–availability gradient in Southern Wisconsin (USA). *Canadian Journal of Forestry Research*, **13**, 12–21.

Nicholls, O. W., Provan, D. J., Cole, M. M. & Tooms, J. S. (1965). Geobotany and geochemistry in mineral exploration in the Dugald River area, Cloncurry District, Australia. *Transactions of the Institute of Mining and Metallurgy*, **74**, 695–799.

Park, D. G. (1982). Seedling demography in quarry habitats. In *Ecology of Quarries*, ed. B. N. K. Davis, pp. 32–4. Cambridge: Institute of Terrestrial Ecology.

Rimmer, D. L. (1982). Soil physical conditions on reclaimed colliery spoil heaps. *Journal of Soil Science*, **33**, 567–79.

Roberts, R. D. & Bradshaw, A. D. (1985). The development of a hydraulic seeding technique for unstable sand slopes. II. Field evaluation. *Journal of Applied Ecology*, **22**, 979–94.

Roberts, R. D., Marrs, R. H. & Bradshaw, A. D. (1980). Ecosystem development on reclaimed china clay wastes. II. Nutrient compartmentation and nitrogen mineralization. *Journal of Applied Ecology*, **17**, 719–26.

Roberts, R. D., Marrs, R. H., Skeffington, R. A. & Bradshaw, A. D. (1981). Ecosystem development on naturally colonized china clay wastes. I. Vegetation changes and overall accumulation of organic matter and nutrients. *Journal of Ecology*, **69**, 153–61.

Schaller, F. W. & Sutton, P. (eds) (1978). *Reclamation of Drastically Disturbed Lands*. Madison: American Society of Agronomy.

Slatyer, R. D. (ed.) (1977). *Dynamic Changes in Terrestrial Ecosystems: Patterns of Change, Techniques for Study and Application to Management*. Paris: UNESCO.

Smith, M. A. (ed.) (1985). *Contaminated Land: Reclamation and Treatment*. New York: Plenum.

Smith, R. A. H. & Bradshaw, A. D. (1979). The use of metal tolerant plant populations for the reclamation of metalliferous wastes. *Journal of Applied Ecology*, **16**, 595–612.

Williamson, N. A., Johnson, M. S. & Bradshaw, A. D. (1982). *Mine Wastes Reclamation*. London: Mining Journal Books.

Woodmansee, R. G. (1979). Factors influencing input and output of nitrogen in grasslands. In *Perspectives in Grassland Ecology*, ed. N. French, pp. 117–34. New York: Springer.

*Virginia M. Kline*
University of Wisconsin-Madison Arboretum and Department of Botany

*Evelyn A. Howell*
Department of Landscape Architecture and Institute for Environmental Studies,
University of Wisconsin-Madison

# 6 | Prairies

Prairie restorations were among the first plant community restorations attempted (Sperry, 1983), and prairie continues to be a community of special interest to restorationists, at least partly because prairies lend themselves to experiments carried out on a small scale, over a reasonable length of time, and even using many conventional agricultural techniques (see Ch. 3). Prairies are planted for different purposes, however, and projects vary greatly in scale. At one extreme are the relatively small, simplified, stylized natural landscapes used in residential and industrial sites to create an esthetic statement of the "visual essence" of native prairies (Morrison, 1975, 1979; Diekelmann & Schuster, 1982); at another are complete community restorations, such as Curtis and Greene Prairies at the University of Wisconsin-Madison Arboretum (Ch. 17), in which the goal is to establish the structure, species composition and interactions characteristic of a functioning prairie ecosystem. Prairie plantings are also used in parks and along roadsides to achieve a relatively open view and provide a variety of color, texture and form, and at the same time minimize maintenance costs (Morrison, 1981); and on sanitary landfills and mine tailings to reduce erosion and provide cover for wildlife.

Prairies are biological communities dominated by grasses and having less than one mature tree per acre (Curtis, 1959). Among the grasses are an array of herbaceous, non-grass-like flowering plants collectively called *forbs*. Although grasses are generally more abundant than forbs, there are often many more forb species than grass species. Prairies vary greatly, however, in species composition and diversity. These differences reflect their world-wide geographic distribution. Prairies are found on most continents, and at one time they represented almost 24 % of the total land cover (Risser *et al.*, 1981). Regional differences can be recognized as grassland "types". In North America, for example, these include tallgrass, mixed-grass and short-grass prairies. Even within a given site, however, climatic fluctuations and other stochastic events can change the nature of a prairie from year to year, causing species to increase or decrease in abundance or even to appear or disappear (Weaver, 1954; Ch. 17).

Because of these regional and temporal variations, it is difficult to make sweeping generalizations about prairie restoration. Within a given prairie region, however, it is possible to be more certain about what is going on. Because most of the published literature concerns prairie restorations in the North American Midwest, and because most of our experience has been gained in this region, our discussion will focus on the tallgrass prairie of the prairie peninsula (Transeau, 1935).

There are two basic approaches to prairie restoration: upgrading an existing degraded prairie (Holtz & Howell, 1983; Vogl, 1964); and establishing the community on sites without existing prairie species (Rock, 1975). Although some of our comments apply to both, most of our discussion pertains to the latter approach.

Each attempt at prairie restoration reported in the literature has its own circumstances of site; of rain and temperature patterns before, during and after planting; and of seed and plant sources. But even though these may lead to unique results, it is possible to make several generalizations about the process. Among these are the fact that fire is an essential tool for the restoration and management of prairies in some areas; and the fact that, even with the use of fire and careful planting techniques, most restored prairies contain unwanted species. Prairie plantings can be established by several planting methods and with several site preparation techniques, and generally require three to five years to take on the appearance of a prairie. Some species are more readily established than others, and some may appear in abundance in the early years of restoration, but later decrease or disappear. Several of these generalizations and others about prairie res-toration and the dynamics of restored prairies are discussed by Cottam in Chapter 17.

### Questions generated by the restoration process

Most efforts are directed at solving two basic groups of problems: increasing populations of less successful prairie species; and getting rid of exotic herbaceous species and woody, usually native, invaders. The search for solutions to these problems leads us to questions that challenge our understanding of a number of ecological concepts.

*Underrepresented prairie species*

Some prairie species have proved to be difficult to establish or to be very slow to spread from their original planted locations (e.g., Sperry, 1983). This failure may be due to unfavorable edaphic conditions or to such biotic interactions as interspecific competition, parasitism and lack of pollinators. It may also involve species attributes such as germination rates, phenology of flowering and fruiting, seed production, capability of vegetative spread and height at time of flowering.

Unfortunately, there have been few life history studies of prairie species, and therefore critical information is lacking, but carefully planned restoration projects do provide opportunities for answering these questions. Using 10 selected "unsuccessful" prairie species, for example, we are initiating a five-year study combining field observations of flowering, seed production and seedling establishment (in disturbed and undisturbed prairie sod) with competition experiments set up as part of a planned 2 ha addition to Curtis Prairie. The study species will be grown alone, with weeds and with more successful prairie species. To determine whether the desired species will compete more successfully after one or two years of growth without competition, the weeds and competing prairie species will also be added to one- and two-year-old pure stands of the study species. We expect the results not only to have practical value, but also to contribute to our understanding of the role of competition in prairie dynamics. Although it is currently customary in restoring prairies to plant all the prairie species the same year, this practice may reduce diversity by selecting for a particular group of species (which are then classified as "successful").

Early successional models also may be tested by varying the site preparation techniques used for new prairie plantings. Currently most restorationists emphasize rigorous site preparation – including either repeated cultivation or herbicide application or both – to discourage weeds. Examples of successful prairies established in this way include those at the International Crane Foundation, Baraboo, Wisconsin; and Fermilab National

Accelerator Laboratories, Batavia, Illinois. This approach is based on at least two assumptions: the first of these is that the initial floristics model (Egler, 1954; Connell & Slatyer, 1977) fits the situation, that is, that species arriving first on a site have a long-lasting advantage; the second is that prairie species do not need the site amelioration afforded by weeds, or that a mix of prairie species will include enough species with pioneering capability to provide any site improvement that is necessary.

In contrast, planting techniques used for the planting of the world's oldest restored prairie, Curtis Prairie, and later the Henry Greene Prairie (also in the University of Wisconsin-Madison Arboretum), were based on a relay floristics model. There was little or no soil preparation since it was expected that the introduced prairie species could displace the agricultural weeds occupying the site, an assumption later modified to credit the important role of fire in the displacement process (Blewett, 1981; Cottam, Ch. 17).

Although both approaches have been successful in producing good stands of prairie grasses and forbs, careful site preparation does seem to speed the process. It would be interesting to know not only whether long-term problems with exotic weeds can be avoided by careful site preparation, but also whether long-term differences in prairie composition and species patterns will be related to the original site preparation.

### Exotic herbs

Almost all restored prairies contain non-native herbaceous species, which differ in their ability to persist as the prairie develops under prescribed burn management. These differences raise some interesting questions about competition. Some, such as quackgrass (*Agropyron repens*), curly dock (*Rumex crispus*) and many annuals and biennials are unable to grow with the prairie species and disappear after several years. Others persist under the developing prairie canopy and are widespread, even though severely suppressed. These include several cool-season grasses, most notably Kentucky bluegrass (*Poa pratensis*). These species are released by disturbance, and are frequently seen in flower on ant hills and along paths even in established prairies. A few species are able to thrive and flower even in well-developed prairie. Among these are two ubiquitous and conspicuous monocarpic species, white sweet clover (*Melilotus alba*) and wild parsnip (*Pastinaca sativa*); as well as a few perennials such as leafy spurge (*Euphorbia esula*) and, in wet places, reed canary grass (*Phalaris arundinacea*).

The persistence and vigor of the short-lived *M. alba* and *P. sativa* suggest the continued presence of patches of bare ground suitable for seedling establishment even in well-developed tallgrass prairie. What is interesting is that these non-native, short-lived species seem better able to use these patches than their short-lived, native counterparts, old field thistle (*Cirsium discolor*) and wild lettuce (*Lactuca canadensis, L. ludoviciana*). In Curtis Prairie in 1971 (the year of the most recent census), these species had frequencies of 8 and 10% respectively. In contrast, *P. sativa* and *M. alba* had frequencies of 34 and 17%. Did the non-native species preempt most of the bare spaces? What makes them so successful? When various burning and mowing treatments were tested for control of *M. alba* (Kline, 1987) the percentage of prairie species increasing, decreasing or staying the same was similar in plots having the same treatment even when one had dense stands of *M. alba* at the start and the other did not. What does this imply about competition between the invaders and the prairie species? Interestingly, knowledge of details of the life history of sweet clover has made it possible to devise burning schedules (Heitlinger, 1975; Kline, 1987) that have resulted in dramatic decreases in this species on Curtis Prairie. (See Gross, Ch. 12, for further discussion of experimental studies of the ecology of some early successional species.)

### Invasion by woody species

Invasion by forest trees and shrubs is a serious problem in most of the native and restored prairies in our area, where the climate and most soils are suitable for deciduous forest (Curtis, 1959). Historically prairies in this area were maintained by fire, but now fire appears to be less effective. This might be due to the nature of today's prescribed burns. These are usually carried out when conditions make burning safe – in contrast to presettlement fires, which frequently must have occurred under fire-hazard conditions. Most prairies today are small, and often only a portion is burned at a time; this may also affect fire quality. Intervals between fires may be too long as well, especially in mesic prairies. Just three growing seasons without fire in a brushy area of Curtis Prairie suppressed grass growth so much that there were insufficient fine fuels to carry a fire.

Frequently prairies have woods along some or all of the periphery. Historically when this situation existed, the boundary was not fixed, but advanced and retreated in response to changes in climate and the frequency and intensity of fire. Today there are few places where this kind of fluctuation can take place, the usual situation being a firebreak separating woods and prairie. In this situation the woods can advance across the firelane by seed or by rhizome, often benefiting from the new conditions produced by the advancing front, but because the fires never go beyond the firebreak the prairie can never advance into the woods. This greatly reduces the effectiveness of fire in maintaining open prairie. For example, tall pines along the southern boundary of Curtis Prairie shade the prairie edge, retarding snowmelt and mitigating both wind and heat in summer. As a result a 20 m strip of prairie lacks sufficient fine fuels to carry prescribed burns well and supports excessive woody invaders.

The change in the species of trees invading the prairie in the past century may be important as well. Before settlement in the mid-nineteenth century, oaks and sometimes aspen were the usual invaders (Cottam, 1949; Curtis, 1959; Muir, 1965). While these species are still important, others have joined them since settlement, including black cherry (*Prunus serotina*) and box elder (*Acer negundo*), native species restricted before settlement at least partly because of fire; and black locust (*Robinia pseudoacacia*), which has been introduced from farther south. Does the forest advance more rapidly because of these new species which diversify the attack? *P. serotina*, for example, a bird-dispersed species, is now a primary invader of dry or mesic prairies. Its fruits attract birds, which bring in fruits of shrubs such as gray dogwood (*Cornus racemosa*), choke cherry (*Prunus virginiana*) and blackberry (*Rubus spp.*), which can start easily in the weakened prairie sod beneath the tree. This whole process may take place in a much shorter time interval than is required for invasion by oaks.

Climate change, an important factor in past vegetation changes (Peters, 1985), may also be affecting woody invasion of prairies today. For the past nine years, in this area, mean annual precipitation has been above normal (Clark, 1986). The timing of precipitation has changed in recent years as well, with more rain occurring in spring and fall, and less in mid-summer. If these changes are encouraging woody invasion, what are the implications for prairie management? How much of a lag is there between climate change and vegetation response?

### Animals in new prairies

The emphasis in creating new prairies has been on the establishment and management of the plant component, the assumption being that creation of prairie habitat will be followed by natural recruitment of animals. We expect that more attempts to introduce animals will take place in the future. Some species of small mammals, birds and insects *do* readily appear in reconstructed prairies. However, some species vital to prairie dynamics may be missing. For example, a prairie may lack mound building ants, which are the major soil cultivators of tallgrass prairie (Baxter & Hole, 1967) and which provide small-scale disturbances that may be important for establishment of certain prairie forbs. Interestingly, most of the ant mounds in Curtis Prairie are in a small area that is the only unplowed natural part of the prairie. In 50 years there has been little spread of ants into the planted area. What keeps the ants from colonizing? What might be done to initiate colonies in a reconstructed prairie? In what way do the soil and vegetation change after introduction of ants, and how quickly do the changes take place? (See Ashby, Ch. 7, for a similar account of missing ants in a restored forest.)

The size of a restored prairie is usually determined by such practical considerations as availablity of land, labor and money, but size may be an important limit to the establishment of some prairie animal species. Large ungulates, such as bison, are obvious examples, but size is important to some prairie birds as well (Sampson, 1983). Furthermore, effective size may be more or less than actual dimensions, depending on adjacent vegetation. For example, if shrubs and trees border a prairie edge, birds such as indigo bunting and gray catbird will be encouraged, but prairie birds such as bobolink and Henslow's sparrow may be discouraged unless the prairie itself is very large. Conversely, can adjacent open land such as farm fields increase the effective size for birds? Upland sandpipers, a species which in Sampson's study was shown to require prairie islands of over 10 ha to breed and of 160 ha to maintain a breeding population regularly, are uncommon in southern Wisconsin, but are frequently seen there on a very small (0.5 ha) natural prairie that happens to be surrounded by cultivated fields. In what ways are these fields ecologically similar to prairies, and in what ways are they different? What would happen, for example, if the grass greenways maintained for erosion control in cultivated fields were planted to prairie? In addition to possible environmental benefits, this would provide a way of testing experimentally existing ideas on the area and habitat requirements of species.

### Long-term dynamics

Many important ecological questions about prairies can be answered only by long-term research. Restored prairies may be very advantageous for this type of research, especially if there are good records of the plant and animal introductions and the management procedures carried out. Some of the questions that might be asked include:

☐ *Disturbances.* How can a long-term dynamic mosaic of disturbance similar to that of presettlement prairies be developed in reconstructed prairies? Introduction of ants has been suggested above to create one type of disturbance, but is it also important to produce artificial bison wallows for large scale disturbance? If so, how would the results be affected by the presence of exotic invaders?

☐ *Population explosions.* Are the observed explosions of rattlesnake master (*Eryngium yuccifolium*) and wild indigo (*Baptisia leucantha*) (Cottam, Ch. 17) in a planted mesic prairie due to the relative youth of the prairie, or are population explosions to be expected over longer time periods as well? What causes them? Can they be predicted? Do these species later decline?

☐ *Local ecotypes.* It is usually recommended that plantings be made with local ecotypes, but what are the long-term consequences of using ecotypes from more distant areas? Will imported ecotypes displace local ecotypes? And, if so, why? Will they outcompete other local species? On the other hand, will the increased genetic diversity represented by the introduction of imported ecotypes benefit the population? Should an effort be made to introduce into restored prairies some ecotypes from warmer or colder areas to anticipate predicted climate changes? (See McNeilly, Ch. 18.)

☐ *Changes in forb frequencies.* Is the tendency noted earlier for more forbs to increase in frequency each year than to decrease or show no change due to the youth of the prairies that have been studied? Will some sort of equilibrium eventually be achieved?

Prairies are especially suitable for the kinds of research suggested by the questions we have raised here as a result of experience with tallgrass prairie restoration in southern Wisconsin. Prairie species mature considerably faster than forest trees, yet many prairie species, like trees, are very long lived. A complex and dynamic community develops rather quickly in prairie restoration, and measurable changes take place on a scale of decades even though it may take centuries to achieve a prairie comparable to the presettlement prairies.

### References

Baxter, F. P. & Hole, F. D. (1967). Ant (*Formica cinerea*) pedoturbation in a prairie soil. *Soil Science Society of America Proceedings*, **31**, 425–8.

Blewett, T. J. (1981). An ordination study of plant species ecology in the Arboretum prairies. PhD dissertation. Madison: University of Wisconsin.

Clark, D. R. (1986). *Annual Precipitation Summary for Wisconsin 1985*. Wisconsin Geological and Natural History Survey.

Connell, J. H. & Slatyer, R. O. (1977). Mechanisms of succession in natural communities and their role in community stability and organization. *American Naturalist*, 3, 1119–44.

Cottam, G. (1949). The phytosociology of an oak woods in southwestern Wisconsin. *Ecology*, 30, 271–87.

Curtis, J. T. (1959). *The Vegetation of Wisconsin*. Madison: University of Wisconsin Press.

Diekelmann, J. & Schuster, R. (1982). *Natural Landscaping*. New York: McGraw-Hill.

Egler, F. E. (1954). Vegetation science concepts. I. Initial floristic composition – a factor in old-field development. *Vegetatio*, 4, 412–7.

Heitlinger, M. (1975). Burning a protected tallgrass prairie to suppress sweetclover, *Melilotus alba*, In *Prairie: A Multiple View*, ed. M. K. Wali, pp. 125–32. University of North Dakota Press.

Holtz, S. L. & Howell, E. A. (1983). Restoration of grassland in a degraded woods using the management techniques of cutting and burning. In *Proceedings of the Eighth North American Prairie Conference*, ed. R. Brewer, pp. 124–9. Kalamazoo: Western Michigan University.

Kline, V. M. (1986). Response of *Melilotus alba* and associated prairie vegetation to seven different burning and mowing treatments. In *Proceedings of the Ninth North American Prairie Conference*, ed. G. K. Clambey and R. H. Pemble, pp. 149–152. Fargo: Tri-College University.

Morrison, D. G. (1975). Restoring the native midwestern landscape. *Landscape Architecture*, 65, 398–403.

Morrison, D. G. (1979). The prairie invades the back yard. *Landscape Architecture*, 69, 141–5.

Morrison, D. G. (1981). Utilization of prairie vegetation on disturbed sites. In *Transportation Research Record 822, Landscape and Environmental Design*, pp. 10–17. Transportation Research Board, National Academy of Sciences.

Muir, J. (1965). *The Story of My Boyhood and Youth*. Madison: University of Wisconsin Press.

Peters, R. L. II. (1985). Global climate change: a challenge for restoration ecology. *Restoration & Management Notes*, 3, 62–7.

Risser, P. G., Birney, E. C., Blocker, H. D., May, S. W., Parton, W. J. & Weins, J. A. (1981). *The True Prairie Ecosystem*. US/IBP Synthesis Series 16. Stroudsburg: Hutchinson Ross Publishing Company.

Rock, H. (1975). *Prairie Propagation Handbook*, 5th edn. Wisconsin: Boerner Botanical Gardens, Whitnall Park, Hales Corner.

Sampson, F. B. (1983). Island biogeography and the conservation of prairie birds. In *Proceedings of the Seventh North American Prairie Conference*, ed. C. L. Kucera, pp. 293–9. Springfield: Southwest Missouri State University.

Sperry, T. M. (1983). Analysis of the University of Wisconsin-Madison prairie restoration project. In *Proceedings of the Eighth North American Prairie Conference*, ed. R. Brewer, pp. 140–7. Kalamazoo: Western Michigan University.

Transeau, E. N. (1935). The prairie peninsula. *Ecology*, 16, 423–37.

Vogl, R. J. (1964). Vegetational history of Crex Meadows, a prairie savanna in northwestern Wisconsin. *American Midland Naturalist*, 72, 157–75.

Weaver, J. (1954). *North American Prairie*. Lincoln: Johnson Publishing Company.

*Joyce Powers*

# Restoration practice raises questions

Restoring or reconstructing plant communities is a relatively new enterprise, and those of us who do it frequently find ourselves encountering questions that challenge the existing ecological descriptions and accounts of the communities we are trying to restore.

For example, ecologists typically describe freshwater wetland plants in terms of the depth of water in which they are found – *Typha* species in 2–12 inches of water, *Scirpus acutus* 1–3 feet, and so forth. This seems straightforward enough, and one might suppose that with this information and the engineering plan for a newly constructed wetland, it would be a fairly easy matter to develop a sensible planting plan for the area.

However, in a recent project of this type, we encountered some interesting complications. For one thing, the precision of the large earth-moving machinery used to grade the wetland substrate turned out to be plus or minus 6 inches, leaving us with a peculiarly uneven planting surface. Even more disconcerting, however, were variations in the texture of the substrate itself. This did not become apparent until we started planting.

After checking out the water depths, we sorted plants accordingly and donned hip boots for planting. In water that measured 2 inches deep before I set foot in it, I sank another 4 inches with the first step, 6 inches with the next, and not at all with the third. It turned out no one had instructed the contractor on the type of soil materials to use. As a result, he graded the site with whatever was closest. Most of the area was organic muck, but there were random patches of sand and clay subsoil as well.

Planting expensive propagules on such a site, one finds oneself asking certain questions: what, for example, is the ecologically effective depth of water on such a site? Is it the same as the nominal depth? Or does it vary with the texture of the substrate? Exactly what are the critical parameters that are related to water depth, and that influence the distribution of plants in a natural wetland?

I have worked extensively with prairie communities, but still find many questions to ask. I've just walked over an exquisite hillside prairie remnant in south-western Wisconsin; what impressed me most about this natural prairie was the nearly uniform distribution of the species over the three-acre site. I have most of the same species growing in my 12-year-old restored prairie, but their arrangement is quite different. *Coreopsis palmata*, for example, grows in clones two feet across in my reconstructed prairie. On the remnant hillside prairie, there are single stems of this plant here and there over most of the prairie, but there are no patches at all. This makes me wonder what we are doing as restorationists that is different from what happens naturally in the making of a prairie? Is it simply the limited time available? Or is it something else?

I've often been told to be careful of certain "aggressive" allelopathic plants – pussytoes (*Antennaria neglecta*), for example. This species formed large monotypic patches on the UW–Madison Arboretum's Curtis Prairie. However, that has not happened on the natural hillside prairie. There, pussytoes grow happily, and non-exclusively, intermingled with birdsfoot violet, star grass, blue-eyed grass, Cynthia, strawberries, and Pennsylvania sedge. One question which needs to be answered is how allelopathy affects the distribution of plants in a community developing over a long period of time.

A related question arises from observing the results of seeding many prairies. From long experience I now know that forb species vary widely in their ability to establish from seed; some establish reliably on nearly every site. Others occasionally establish from a seeding in the initial stages of restoration, but cannot be counted on to do so. And still others never grow from an initial seeding, even though they are easily grown in a nursery.

It is now common practice for our company to seed the forbs in successive waves, and in this way we have been able to establish many of the more difficult species. However, we do not really know what is going on here – so we have more questions. What mechanisms are responsible for these patterns of success and failure? What are the factors determining survival, succession, and plant distribution in these communities?

*W. Clark Ashby*

Department of Botany, Southern Illinois University

# 7 | Forests

The restoration of forests differs in scale – time, breadth, height and depth – from the restoration of most other types of vegetation. It has been practised by planting trees for many years and in many parts of the world. The possibilities of full-scale forest restoration are, however, only now being realized. One of its major rewards will be better understanding of the ecology of natural forests, which are relatively intractable experimental subjects for small-scale projects. This chapter will explore various types of restoration projects, with an emphasis on midwestern deciduous forest types, and on what we might expect to learn as a result of attempts to restore them.

Forest restoration will be more fruitful if we have a clear concept of what constitutes the essence of a forest. In turn, restoration studies may be a chief means to develop such a detailed concept. On a broad formation or biome scale, the forests of the United States seem reasonably distinct and the factors influencing their distribution fairly clear. Postulated endpoints of forest development are based on the concept that long-term vegetation (Major, 1951) and soil development (Jenny, 1941, 1980) are both dependent variables reflecting the independent variables of climate, parent soil material, relief or topography, and available organisms. Thus, the boreal forest and the deciduous forest have numerous distinguishing features, and reasonably distinctive ranges within which each is accepted as the normal product of natural forest development.

Variations within the theme of boreal forest and of deciduous forest are well known, however, and it is at this level of community organization that an uncertainty principle perplexes ecologists. The element of predictability is largely lost within smaller-scale forest units, the scale at which autecology and synecology may seem to merge. Significantly, this is also the feasible scale for restoration experiments.

Forests are built both naturally and artificially by successively adding species until a self-sustaining entity is created. Restoration experiments can test to what degree the whole is greater than its parts. In what ways are component species similar and dissimilar? Can the distribution limits of a forest type be equated with or somehow related to the limits of its component species?

There is a paradox in autecological studies. Reciprocal transplant, or provenance, studies (themselves a form of restoration ecology) have revealed previously unsuspected genetic tailoring of locally adapted populations (ecotypes) to growing conditions within the total expanse of those forests in which a given species is found. Despite these evident local adaptations, however, forest trees have been planted far beyond their native ranges in forestry and in horticultural practice. Sometimes they are marginally successful and sometimes enormously so – Monterey pine (*Pinus radiata*), for example, grows much better in New Zealand and Australia than in its limited native range in California. Similarly, southern pines, shortleaf (*P. echinata*) and loblolly (*P. taeda*), have been widely planted in the lower midwest.

At the same time, experiences with such tree plantings suggest that inherent genetic variety is ecologically significant. Attempts by arboreta and botanical gardens to transplant species frequently meet with difficulty. Even when successfully established in a community closely resembling that to which they are native, these misfit genomes rarely persist in a new locale without continual management or artificial disturbance.

This, however, is only a single example of the subtle adaptations and interactions within a community which may be uncovered and characterized through attempts at restoration or partial restoration of the community. The point is that properly designed restoration experiments offer almost unlimited opportunities for elucidating these factors in the ecology of the forest community or ecosystem. To be most productive, however, these experiments in restoration ecology should be designed specifically to test well defined hypotheses. The following are discussions of a number of hypotheses related to forest structure, dynamics, and functioning which might be tested in this manner.

☐ Some typical stages of forest development are not essential and may be bypassed.

☐ Forests require a minimum area and number of components.

☐ Factors in the development of midwestern forests include:
  1. Seed dispersal, competition, predation and stochastic events
  2. Stress and environmental change
  3. Depletion of genetic resources
  4. Exotic species.

### Some typical stages of forest development are not essential

Classical succession theory suggests that, within the natural range of a forest type, change in a community occurs as the existing plants, animals, and microorganisms alter the conditions within the community, so that many or all of the existing species are eventually replaced by others. Thus, both the species composition and the physical structure of the community typically change over time.

While succession theory has been central to ecological thinking for nearly a century, many questions remain about the nature of the process, the extent to which it actually occurs in particular communities under specific conditions, and also about the detailed mechanisms underlying reported examples of its occurrence. Postulates of predictability in succession, and especially of the climax status of vegetation, have been widely challenged.

Studies of succession have to date been mainly descriptive. Moreover, since succession, especially in its later stages, tends to occur extremely slowly, investigators have usually attempted to characterize the process by studying variations in communities across space on the assumption that this provides a cross-section of a process actually occurring in time. Classic examples of such studies have been studies of the vegetation invading old fields or small lakes. Such studies may have limited value, however, because of possible unknown earlier difference in climate, species availability and other complicating factors. Similarly, questions exist about the various mechanisms proposed to account for changes in communities. Since these have generally been adduced from descriptive studies, their actual role in community change is often not clear. In many cases they could almost certainly be clarified by manipulative experiments.

This brings us directly to the subject of restoration ecology, or the restoration of communities planned and carried out specifically to test ideas about how they develop and function. Since active restoration is to a very large extent a process of bringing about changes that would occur naturally given time, but causing them to occur much more rapidly, restoration may be viewed as an attempt to imitate succession in order to control it. Thus the attempted active restoration of a community naturally draws attention to critical factors in the process – especially those that have been neglected or improperly handled – and provides countless opportunities to identify them and to determine more precisely the role they play in community change.

In other words, restoration, especially when carried out in the context of restoration ecology, amounts to the experimental testing of the ideas underlying succession theory.

This is clear, moreover, from the record of restoration efforts, despite the fact that these have often not been carefully planned and controlled experiments, but have been trial-and-error projects carried out for more or less immediately practical purposes. A basic tenet of succession theory, for example, is that certain species serve as pioneers and give way to succeeding species as site conditions and competitive relations are altered. Virtually no experiments, however, have been carried out specifically to determine the direction of change or the forces influencing it. Under these circumstances, forest restoration projects can provide some of the best information about this process, even though they may not have been set up specifically as experiments to test ideas about succession (Ashby, 1984).

In the case of hardwood forests in the lower midwest, for example, sassafras (*Sassafras albidum*) and persimmon (*Diospyros virginiana*) have typically been considered to be early successional species that presumably will be succeeded by oaks and other late-successional hardwoods (Drew, 1942; Bazzaz, 1968). This concept is based almost entirely on observations of existing communities differing in age. With better soil conditions, succession to tulip tree (*Liriodendron tulipifera*) or sugar maple (*Acer saccharum*) as the first tree stage have also been found (Ashby & Weaver, 1970). On the basis of the majority of cases it might be inferred that sassafras and persimmon were a necessary stage in community development and that they play some highly specific role in succession on these sites. Experience with the rest of these communities, however, clearly shows that this is not the case. Because of the limited economic value of sassafras and persimmon, and also because it is often difficult to obtain and handle stock of these species, they are rarely planted on disturbed sites in this area, various southern pine species often being used instead. As it happens, these pines, once established, are so readily invaded by the "later successional" hardwoods that many foresters consider that the best way to grow hardwoods in this area is to plant pines.

This experience obviously has considerable practical value. At the same time it has provided clues to the factors influencing vegetation change in these communities. For one thing, it makes it clear that a specific association, the sassafras-persimmon association, is not essential to later invasion by hardwoods, but is essentially replaceable by another, quite different kind of community. At the same time, it is clear that the nature of the pioneer community *does* influence the more mature community that succeeds it. Whereas succession has proceeded in artificial communities quite unlike the naturally occurring pioneer communities of this region, it has not proceeded in exactly the same ways. For example, whereas native sassafras-persimmon stands are generally invaded by oaks, pines are usually invaded both by old-field species such as sassafras and by later successional oaks. In contrast, planted stands of black locust (*Robinia pseudoacacia*) tend to be invaded by elm (*Ulmus americana*), black cherry (*Prunus serotina*), boxelder (*Acer negundo*), hackberry (*Celtis occidentalis*) and other mesic forest species (Ashby, 1964).

These observations naturally lead to questions about the nature of the factors influencing succession and the composition of the succeeding community. Pines, for example, create shade, alter water budgets, increase soil porosity and build organic matter as black locust does, but, unlike locust, they maintain a year-round canopy and have relatively little influence on the level of soil nitrogen. Clearly, restoration efforts designed specifically as experiments to isolate these factors and determine the role they play in succession are needed now.

The potential value of such studies is evident from the results of four plantings on an abandoned field in southern Indiana, reported by Carmean *et al.* (1976). Height growth at 16 years of planted black walnut (*Juglans nigra*), sweetgum (*Liquidambar styraciflua*) and tulip tree was greatest on the part of the field clearcut from a black locust plantation, and least on the portion occupied by broomsedge (*Andropogon virginicus*), poverty grass (*Danthonia spicata*) and goldenrod (*Solidago* spp.) at the time of planting. Heights were intermediate and approximately equal on plots clearcut from a shortleaf pine plantation and on a portion of the field fully stocked with sassafras, persimmon and elm. Planted red oak (*Quercus rubra*) grew best on plots clearcut from black locust and shortleaf pine, relatively poorly on the formerly fully stocked area, and, like the black walnut, did not live past ten years on the non-stocked, herbaceous area of the field. These growth results were related to foliar nitrogen levels, noncapillary soil pore volume, bulk density, and organic matter content, and to mycorrhizal and phosphorus fertilization studies carried out in pots. Allelopathic compounds produced by broomsedge and sassafras were also considered to have influenced the growth results.

The findings of this study showed clearly how antecedent vegetation influences both environmental conditions and growth of preclimax to climax species. It also helped identify some of the factors involved. Restoration experiments focusing on pioneer, intermediate, and late-successional species can similarly identify their needs and roles in succession.

Succession theory predicts that late-successional species will tend to fail when introduced into a disturbed or open site. In fact, this has generally been the case. Late-successional hardwoods such as white oak (*Q. alba*) have generally failed when introduced directly into open sites in the midwest. Observations of these projects, moreover, have suggested that these failures have generally been due to unfavorable water and nutrient relations, or occasionally to animal damage – in other words to unsuitable conditions that might have been meliorated by pioneer species or, once the critical factors are understood, by management procedures.

There have been exceptions, however. Hardwoods have succeeded in tree plantings on initially bare land surface mined for coal in many parts of the country (Ashby, Kolar & Rogers, 1980; Larson & Vimmerstedt, 1983), and it is these cases, even more than those in which failure occurred as expected, which have been most useful in refining ideas about the extent to which succession actually occurs in these communities, and the factors influencing it. For example, the pattern of successes and failures and the changes which have taken place in these communities raise interesting questions about the nature of the factors guiding succession.

Some stands of late-successional species such as red, white, and bur oak (*Q. macrocarpa*), as well as red maple (*Acer rubrum*) and silver maple (*A. saccharinum*) and black walnut, have survived on reclaimed land and have also maintained their composition. Others such as sweetgum and ash (*Fraxinus americana*) and tulip tree are now rapidly being invaded by oaks and other trees (Ashby, Rogers & Kolar, 1981; Larson, 1984). Substantially more invasion, by a greater diversity of species, is found in black locust plantings. Obvious questions are: why do certain species succeed on open sites while others almost invariably fail? And why are some communities rapidly invaded while others seem to be extremely stable?

These questions are raised still more forcefully by the observation that in some instances even climax species such as sugar maple have succeeded when planted on open or disturbed sites, suggesting that factors influencing the success of these species are not yet fully understood. One such instance is a mixed, first-tree-generation stand of sugar maple and other hardwoods planted on mined land near Terre Haute, Indiana, which is growing well, reproducing, and gives every indication of successful establishment. This stand lacks nearly all the usual associates of sugar maple, and has bypassed perhaps four successional stages which typically precede sugar maple stands in that part of Indiana. Two obvious factors in this development are man's early introduction of seedling maples to the site, and deep, porous, well-drained soils with coarse fragments that weather to release mineral nutrients. Equivalent soils take many years to develop on old, eroded fields, if indeed they ever do.

Another factor, discussed by John Aber (Ch. 16) on the basis of studies of planted sugar maple forests at the University of Wisconsin–Madison Arboretum, is that maple has a relatively flat response to increasing supplies of nitrogen. Fresh minesoils typically are low in decomposable organic matter and in available nitrogen. Maple establishment on old-field sites in the Arboretum may also reflect its low nitrogen demands.

In general, the results of these *ex post facto* "experiments" in restoration ecology suggest that the forces guiding succession are a good deal less specific than might have been expected, so that in at least one instance, often used as a classic example of succession, the process is actually so loose or tenuous that it is relatively easy to skip whole stages entirely. So far, observations such as those described above have not been integrated into succession theory. Carefully designed restoration experiments appear to offer the best available means to do this, and in the process to explore the interactions between forest species and between species and the changing environment of the forest.

### Forests require a minimum number of components and area

Four criteria for recognition of a forest type are physiognomy, floristic composition, ecological structure, and habitat relations. These criteria are useful for analysis of forest development. The dominant plants and animals in the forest communities of the northern temperate zone in the western world have been well described. Appreciable information is available for selected fungi, arthropods, and annelids. Temperate deciduous forests have perhaps 1000 species of higher plants, many kinds of higher animals, and perhaps countless kinds of invertebrates, lower plants, and microorganisms.

Although the well-studied mycorrhizal fungi and nitrogen-fixing bacteria and actinomycetes illustrate the importance of microorganisms in community interactions, those interactions are still not well understood (see Miller, Ch. 14). Absence of this kind of information renders predictions of community development very tentative. Even less is known about many other groups. Experimentation to learn how or if these organisms can reach a restored forest is almost a necessity, for deliberately introducing them is usually not practical. Consequences of the absence of one or more groups, if recognized, would shed valuable light on "normal" forest functioning.

The very concept of restoration ecology may imply that present communities are reasonably fully stocked and that the major task is to replicate that assemblage of species. Missing components are, however, characteristic of several kinds of forests. An increase in environmental stress is commonly assumed to bring about a decrease in community floristic and structural complexity, as illustrated by river floodplain forests. There are, however, other, unexplained, missing components in forests. Mesic forests in southern Illinois lack the prominent viburnums and other shrubs so characteristic of more northern and eastern deciduous forests. Do these missing components represent unfilled niches, or are they the result of unrecognized stress factors? Several types of plantings, or restoration ecology experiments, could be used to differentiate unfilled niches from limitations related to environmental stress. Exotic species, discussed later, are already furnishing some of these answers.

Knowledge of how – or even whether – most species reach and disperse into a forest is commonly lacking. Orchids – but not violets – are often found in forests on lands stripmined for coal. One might suppose that the chance introduction or the planting of a few individuals would suffice to introduce a species into a community, but this may not always be true. Experience to date in natural succession and various types of plantings is that dispersal of many species remains spotty, even following introduction. The major limiting factor seems to be dispersal agents.

Forest plantation monocultures have shown the effectiveness of birds in distributing trees with fleshy fruits such as black cherry. Some of the small – seeded oaks (for example, pin oak, *Quercus palustris*, and black oak,

*Q. velutina*) are carried and cached in trees by bluejays and crows, and may be dispersed in this way. Persimmons are brought into new areas by raccoons and foxes. Hickories (*Carya* spp.), walnuts, and red and bur oaks typically await the entry of squirrels into an area before these large-seeded trees make an appearance.

An example of a missing dispersal agent was the failure of bloodroot (*Sanguinaria canadensis*) and wild ginger (*Asarum canadense*) to disperse over a period of years from plantings along paths in a restored sugar maple community at the University of Wisconsin–Madison Arboretum. Recent research (Woods, 1984) strongly suggests that this is due to the absence of certain species of ants, which normally disperse seed of these species to new growth sites. Thus, restoration here led to a clearer idea of the role of a single element that had been omitted in the restoration effort.

Other considerations apply to floristic and faunistic composition, with numerous complications. A desired species may have specialized, often unknown, requirements for long-distance propagule transport and for establishment. Animals may prevent its establishment unless it is protected for several years. All of these factors can be tested in the course of forest restoration and applied to rehabilitation of partially disturbed forests. Sassafras, for example, typically plays a minor role in forests, but may play a major role in succession. In areas with a high density of deer, sassafras may be eliminated. Restoring sassafras in such areas may yield unexpected dividends in assessing its role in forest functioning.

While scale is widely recognized as an important factor in community development (see Ch. 19), practically no guidelines are available on the minimum area required for the establishment of natural forest communities. Strips of planted trees along highways are not considered forests, although the aggregate area may be large. Edge effects are well known and result in transitional community features. "How small is too small?" is a significant question, and grows in practical importance as available land for forests becomes scarcer. The "island effect", apparent in scattered woodlot remnants of the presettlement forest, is strongly related to size (Burgess & Sharpe, 1981). By restoring communities of different sizes we can test for both minimum size and minimum number of components to sustain a viable forest community.

Ecological structure is a sensitive indicator of local community types. Assuming that the number of strata may vary from one to four or five in deciduous forests, depending on the stress factors affecting a community, necessary components could include canopy trees, understory trees, shrubs, herbs and a thalloid layer. How to get a requisite number of strata is not known. If all are planted at once, the forest herbs and thalloid species will probably die in the absence of a canopy, and the shrubs may overgrow the trees. If trees are planted and establish a canopy, much time may elapse before the other usual structural elements may invade. Exotic plants are also likely to enter the stand and may be very difficult to eradicate. All this raises numerous questions about how forests naturally achieve their more or less complex structure.

It also suggests ways of answering them. Restoration plantings may have to be carried out in stages and handled carefully. Some kinds of shrubs are found chiefly in openings. A young tree canopy may not have openings, however, and decisions will be needed on where to place them, how large they should be, and when they should be created. The opportunities for experimentation and testing hypotheses about factors in forest development are many and varied. For example, young woody plants of all types are susceptible to damage from deer and other animals, whose numbers are also strongly affected by forest openings. This balance of numbers at different trophic levels, which may be decisive in community restoration, can be explored in the course of restoration experiments.

Mature trees require, and physically modify, relatively large locales in which other species are found. If an old-growth forest has 250 trees per ha and 25 constituent tree species (even though unequally represented) many of these species will have unacceptably small numbers in restoration projects of only a few hectares. Partly this relates to pollination, seed dispersal and other problems of maintenance of species in low numbers. A major factor is the adequacy of habitat diversity in small areas. The success or failure of a restored forest often varies from area to area, and even from tree to tree. Properly interpreted, this may provide information on the complexity of total community needs, and on species requirements and interactions. This, in turn, may help us understand why, though it may be easy to grow component species separately – in plantations, for example – it may be much more difficult to achieve realism and permanency in restoring or transplanting entire forest communities. In general, replication of suitable growing conditions for establishing a single species is much easier, and depends much less on chance events, than replication of the diverse conditions of moisture, aeration, nutrients, light and other factors that occur within a forest throughout its life.

### Factors in the development of midwestern forests

*Seed dispersal, competition, predation and stochastic events*

This discussion will focus on these factors related to oaks. The vegetation of much of the Central Mississippi Valley is classified as some type of oak forest. Oaks are rarely in the first wave of tree invasion in succession, though they usually regenerate well after logging, and oak forests are managed with this in mind. Why oak invasion is delayed in other types of succession is not known, however, and attempts to introduce these species in reclamation and restoration projects have provided some clues. In general, the few plantings that have been tried on open sites have failed, but there have been occasional, very informative successes. For example, bur, red and white oaks have often grown well when planted on land surface mined for coal (Fig. 7.1), and the success of recent plantings of these species has turned out to correlate closely with the absence of herbaceous cover. Both the plant competition and the high populations of rodents typically associated with dense herbaceous cover are considered to be limiting factors in oak establishment. These observations could be checked further by restoration experiments in which competition and/or animal populations are selectively controlled.

In either case, the lag period for oaks to enter in succession could reflect a need for pioneer trees to shade out the early successional stages of old-field herbs, reducing competition and rodent habitat and increasing the effectiveness of rodent predators. Or might it be due to a need for perches for the birds needed to bring acorns to the area? This type of question is fundamental to forest development and could be answered by restoration experiments with artificial perches to increase predation and encourage acorn dispersal.

"What is the status of oak in our forests likely to be in the future?" is a longer-term question. Oak has long been considered a major component of eastern deciduous forests. Braun (1950) recognized the oak–pine, oak–chestnut and oak–hickory associations. The chestnut (*Castanea dentata*) is now gone as a forest dominant. Are the oaks destined to follow? Throughout the Central Mississippi Valley, oak forests are rapidly being invaded by sugar maple and other relatively mesic species. Instances are known in Illinois, Iowa, and at the University of Wisconsin–Madison Arboretum (McCune & Cottam, 1985), in which oak forests are being taken over by the same combination of black cherry, boxelder and elm, noted earlier as invading

*Fig. 7.1.* Except for the limited representation of oak forest species, this white oak stand closely resembles a natural forest 32 years after planting on fresh mine soil in south-eastern Illinois. Although white oak is not usually regarded as a pioneer species, the success of restoration efforts like this one shows that it can succeed on an open site, if herbaceous cover is not dense. Further restoration experiments might help determine which of several factors are responsible for the failure of oaks on densely vegetated sites (photograph courtesy of W. Clark Ashby).

black locust plantings. The widespread absence of oak regeneration in existing stands may be a prelude to numerous instances of such invasion. What may replace the cherry–boxelder–elm combination is a major question in forest restoration.

Are we seeing the crest of a wave of oak regeneration set in motion by earlier forces no longer operative? Present ages of the dominant oaks in stands that have been studied in southern Illinois are typically about 160 years or less. Postulates on the nature of the earlier conditions favoring oak establishment range from a dramatic decrease in the frequency of fire in the midwest at the time of European settlement to the knocking down of previous forests by the great central Mississippi River Valley earthquakes of 1811–12. Would restoration efforts focusing on oak forests be comparable to the Maginot Line thinking in Europe prior to World War II – a preparation for the last war? Restoration plantings that include various combinations of oaks and other potential climax species could help answer this question, and would also have important implications for the management of recently acquired natural areas in midwestern states.

In general, restoration efforts provide excellent opportunities for answering numerous questions related to oak forests. Those suggested by these discussions include:

☐ What ingredients are necessary for successful oak forest establishment in a successional sequence?

☐ What are the dispersal mechanisms for the trees and other components of an oak forest? How effective are they? What factors influence their effectiveness?

☐ Are there unfilled niches in existing oak forests, as suggested by invasions of buckthorn (*Rhamnus cathartica*) and bush honeysuckle (*Lonicera tatarica*) in the upper midwest, and of Japanese honeysuckle (*L. japonica*) and autumn olive (*Elaeagnus umbellata*) in the lower midwest? If these niches were filled could oaks regenerate? Different planting mixes could test these possibilities.

☐ Are present oak forests pre-climax? Again this could be tested by planting diverse species mixes, including representatives of postulated climax species other than oak.

☐ What other types of species replacement for oaks will be available in the future for midwestern forests if sugar maple does not fill this role?

## Stress and environmental change

The old-growth forests of today developed in environments that differed in many ways from those of the present. Upland habitats are commonly more xeric now than in presettlement times because of drainage-system entrenchment and accelerated runoff. Lowlands have experienced sedimentation, and many areas have a greater incidence of flooding because of levees and dams built to protect agricultural lands. Soil types have changed through erosion, deposition and altered hydrologic conditions.

These environmental changes will influence the types of forest that may develop – or can be restored – in a given locality. Seemingly small shifts in environment can have large effects on community composition because of the high sensitivity of tree species to soil conditions.

Two currently well-publicized changes in environmental quality are changes in the nutrient content and acidity of precipitation. Elevated nitrate levels have been reported to lead to extended growth periods, delayed dormancy and frost damage, and to magnesium deficiency (Kiester, 1985). Community composition will in time reflect the response of tree and other species to new nutrient or acidity levels. Forest restoration studies under artificial stress, or natural stress of limited occurrence, could help identify specifically the impacts of stress factors, to anticipate them, and perhaps to suggest ways of alleviating new types of stress, such as those attributed to acid rain.

A much less recognized, though widespread, stress factor affecting forests and forest restoration is soil compaction. Almost all types of disturbed soils typically have compacted layers – plow pans and disk pans from farming, and traffic pans from logging or from soil replacement in reclamation projects. Such compacted soils have been associated with arrested forest development because they limit the kinds of trees that can grow on a site, and may also restrict the growth of those that do occur. Postulated causes of poor tree growth under these conditions are: limited root-system development due to excessive soil strength in the compacted layers; seasonally water-saturated soils leading to root death; and also exacerbated water stress during drought periods.

Undisturbed soils have fragipans and clay pans, which also affect the distribution and growth of different species in different ways. Pin oaks, for example, are found on tight soils, while reasonably vigorous black walnuts usually occur only on deep well-drained soils. These ecological variables are not generally recognized, and their influence on a community may be attributed to other environmental effects, but they are often extremely important. The occurrence of soil layers that restrict rooting clearly has important effects at the landscape level. South-eastern Illinois, with Grantsburg fragipan soils, for example, has different successional sequences (specifically, a delayed sassafras-persimmon stage) and more xeric old-growth forests than south-western Illinois, with its well-drained Alford soils.

Studies by my research group have shown that compacted soils may severely limit tree growth on lands surface mined for coal. Some effects did not show up for more than twenty years, however, after the trees had made significant growth, and tree species responded very differently to compaction. Oaks, though they may grow less well on compacted than uncompacted sites, are typically among the more successful species. This, of course, illustrates the significance of questions about how impermeable soil layers may influence the distribution of species in the landscape.

Non-pan soils are especially adversely affected by traffic from heavy machinery, and this type of disturbance of the soil/plant system helps uncover factors important to it. Restoration plantings on differentially compacted soils can lead to important insights into the ways this factor affects forest development even under natural conditions, and research of this kind is making it clear that these factors can be as important to the development of the community as the chemical factors that have attracted most of the attention of researchers in the past.

This research has also led to questions about factors that naturally cause or reduce compaction. Restoration studies offer unique opportunities to identify the ways in which compaction is alleviated under natural conditions, and ways a community may respond to improvement in the physical condition of the soil. In some regions freezing and thawing or wetting and drying contribute to the breaking up of soil layers. Much of the potential for reducing stress from compaction can, however, be found in the plants and animals of the community. They can be used as tools to examine the effects of changes in soil strength and bulk density, hydrology, aeration, and also plant responses to these changes. Trees under favorable physical soil conditions have massive root systems. Individual roots decay to provide channels of various sizes, and it is now clear that such "macrochannels" play an extremely important role in forest ecology by influencing the subsurface movement of water and thus the hydrologic balance of a forested area. They also affect soil aeration, movement of soil animals, and subsequent root growth. Early successional or restored forests typically lack these macro-channels, and so offer opportunities to study their role by creating artificial channels and assessing their effects. Posts could be driven through compacted layers and allowed to rot, or shales, which weather readily, could be incorporated into the rooting medium.

Similarly, experiments involving the creation of macrochannels like those created by burrowing animals would be useful. In semi-arid steppes these are recognizable as crotovinas, channels of black surface soil deep in the light-colored parent material (Russell, 1973). Burrows of groundhogs and other mammals are often found in mature forests. What role might they play in shifts of forest tree composition from pioneer to climax?

Smaller channels, also of great importance to good growth of many trees, are formed by the death of smaller roots and by the activities of many soil animals. Earthworms have been well studied and shown to be a major earth-moving organism in forest soils. Mound-building ants penetrate deeply into the soil. Restoration studies with these organisms can illuminate ways in which forest communities are integrated as functional units. Suitable food and environments for their activities must be given to these animal populations, and learning how to do this would be another valuable component of restoration ecology studies. The entry of soil animals into a forest community may be a significant feature of succession, and we need to know what effects can be associated with which species.

The role of porosity and macrochannels in forest restoration could be tested in several ways. One would be to cut off an old-growth stand with as little compaction as possible. A companion area would be logged with extensive equipment traffic to give compaction throughout the area. Suitability for trees and associated forest species in the two areas would then be tested by growing the presumed climax species, such as basswood (*Tilia americana*), which are associated with well-drained soils, climax species such as white oak, which are usually found on shallower soils, and pioneer species such as persimmon, more typical of disturbed upland sites.

### Depletion of genetic resources

Old-growth forests contained allotments of and had continuity with vast gene pools. Today these gene pools have been fragmented by clearing and destruction of forests, with consequences as yet unknown. Restoration studies can test the significance of impoverished genetic potential in local forests by differential use of genetic resources for a species. The plant material used could be taken from different types of habitat within an area, and/or from different geographic locations. Many species of trees have extensive genetic variability of both types, as shown by Plass (1973) for Virginia pine. (For further discussion see Ch. 18.)

Another genetic consequence of habitat disturbance may be the breakdown of genetic barriers, or hybridization. Hybrids are found in many kinds of plants, and both affect and are affected by the forests of an area. An enormous post-settlement variability in *Crataegus* has been attributed to "hybridization" of the habitat. And oaks that do not fit local keys are rapidly invading older, tree-covered stripmines in southern Illinois.

Still another virtually unknown influence in traditional forest ecology is that of vegetative reproduction, or clonal plant populations. Hundreds of quaking aspen trees in an area may belong to a very few, physiologically integrated clones (Cook, 1983). Still another type of physiological integration results from the common occurrence of root grafting in many kinds of trees. The ecological significance of physiologically integrated species groupings could be tested in forests restored with differing types of clonal material, by deliberately altering the extent of root grafting. In general, restoration projects draw attention both to the importance of genetic material and to ecotypic variations, and represent a step toward the more "ecological" taxonomy suggested by Harper (1982) (see also Chs 3 and 21).

### Exotic species

*The old order changeth, yielding place to new*

Whereas 25 years ago the broomsedge, sumac (*Rhus copallina*) and sassafras-persimmon stages of old-field succession were commonplace in southern Illinois, today stands of sericea lespedeza (*Lespedeza cuneata*), mats of Japanese honeysuckle, clumps of multiflora rose (*Rosa multiflora*) and thickets of autumn olive are more likely to meet the eye, especially near urban areas. Introduced woody species have also moved into disturbed forests. Evidently there were to some extent unfilled niches in the several stages of native forest succession.

Although aggressive shrubs have generally played a minor role in the development of our forests, restoration experiments with and without the exotic newcomers could help explain processes important in the development of existing patterns of vegetation. They might also provide insights into future successional pathways that have become problematic in the presence of these aggressive invaders. Savanna-like mixtures of shrubby black locust and tall fescue grass (*Festuca arundinacea*) have been stable for 20 years or more on roadsides and stripmines. Whether or not traditional forest succession can compete with the newly arisen vegetation types and species, the invaders represent probes that help characterize the integrity of the community. Restoration experiments with controlled addition and sub-traction of these species from the community would be an effective use of these probes.

Pioneer vegetation is often marked by localized patches of one or relatively few species. In the absence of competition, the carrying capacity for and populations of a given plant may be much larger than later in succession, as the community becomes more complex and its niches are more fully occupied. This same concept probably also applies to all types of forest organisms. Communities of varying degrees of complexity could be created to test for the impact of differing levels of competition on niche structure and function.

### Conclusion

Overall, restoration bids fair to help us learn much of what is worth knowing about the life requirements of individual species, and how two or perhaps more species interact. It can clarify the community functions of energy flow, nutrient cycling and population regulation. It has already taught us much about succession in forest communities. And it is now possible in many cases to create a stand of trees with a designated species composition. A logical way to proceed further would be to identify conditions needed for each of the cadre of species judged essential for a chosen forest type, create those conditions, plant the species, and await the desired result. This may be naive, however, and so may lead to further lessons. A potential problem is that optimum growing conditions will probably differ for species even within a single forest type, and that the several kinds of trees, shrubs, herbs, microbes and animals will each alter those conditions once the community is launched. How many components are required to restore the natural forest? This will not be an easy question to answer, and successful restoration projects will probably be aided by invasions of needed com-ponents from nearby established communities, further complicating the study.

Restoring forests is a major commitment. Provision must be made for the ultimate massive size and long life of trees. A sufficiently large planting area is needed to get an adequate number of individuals and variety of species to represent a selected forest type and to minimize edge effects. The human element will be of continuing importance because no one individual can handle the ongoing job, and institutional arrangements have had a high failure rate.

Indeed, the greatest obstacle to successful, long-term forest restoration may well be assuring continuance of the project. Legislation similar to that for preserving old-growth natural areas would certainly help encourage forest restoration projects. In general, some kind of pre-established institutionalized arrangement is almost a necessity for long-term forest restoration projects, and the biologist should be aware of these needs when starting a restoration program.

### References

Ashby, W. C. (1964). Vegetation development on a strip–mined area in southern Illinois. *Transactions of the Illinois State Academy of Science*, **57**, 78–83.

Ashby, W. C. (1984). Plant succession on abandoned mine lands in the eastern US. In *Proceedings of the National Symposium and Workshops on Abandoned Mine Land Reclamation*, pp. 613–31. Northwood: Science Reviews Ltd.

Ashby, W. C., Kolar, C. A. & Rogers, N. F. (1980). Results of 30-year-old plantations on surface mines in the central states. In *Trees for Reclamation*, pp. 99–107. USDA Forest Service General Technical Report NE–61.

Ashby, W. C., Rogers, N. F. & Kolar, C. A. (1981). Forest tree invasion and diversity on stripmines. In *Central Hardwood Forest Conference III*, pp. 273–81. Columbia: University of Missouri.

Ashby, W. C. & Weaver, G. T. (1970). Forest regeneration on two old fields in southwestern Illinois. *American Midland Naturalist*, **84**, 90–104.

Bazzaz, F. A. L. (1968). Succession on abandoned fields in the Shawnee Hills, southern Illinois. *Ecology*, **49**, 924–36.

Braun, E. L. (1950). *Deciduous Forests of Eastern North America*. Toronto: Blakiston.

Burgess, R. L. & Sharpe, D. M. (eds) (1981). *Forest Island Dynamics in Man-Dominated Landscapes*. New York: Springer-Verlag.

Carmean, W. H., Clark, F. B., Williams, R. D. & Hannah, P. R. (1976). *Hardwoods Planted in Old Fields Favored by Prior Tree Cover*. USDA Forest Service Research Paper NC–134.

Cook, R. E. (1983). Clonal plant populations. *American Scientist*, **71**, 244–53.

Drew, W. B. (1942). *The Revegetation of Abandoned Cropland in the Cedar Creek Area, Boone and Callaway Counties, Missouri*. Columbia: University of Missouri Agricultural Experiment Station Research Bulletin 344.

Harper, J. L. (1982). *After Description*. Special Publication No. 1, British Ecological Society. Oxford: Blackwell Scientific Publications.

Jenny, H. (1941). *Factors of Soil Formation*. New York: McGraw Hill.

Jenny, H. (1980). *The Soil Resource Origin and Behavior*. New York: Springer-Verlag.

Kiester, E. Jr. (1985). A deathly spell is hovering over the Black Forest. *Smithsonian*, **16**, 211–28 (in part).

Larson, M. M. (1984). *Invasion of Volunteer Tree Species on Stripmine Plantations in East-Central Ohio*. Wooster: Ohio Agricultural Research and Development Center Research Bulletin 1158.

Larson, M. M. & Vimmerstedt, J. P. (1983). *Evaluation of 30–Year–Old Plantations on Stripmined Land in East Central Ohio*. Wooster: Ohio Agricultural Research and Development Center Research Bulletin 1149.

Major, J. (1951). A functional, factorial approach to plant ecology. *Ecology*, **32**, 392–412.

McCune, B. & Cottam, G. (1985). The successional status of a southern Wisconsin oak woods. *Ecology*, **66**, 1270–8.

Plass, W. T. (1973). Genetic variability in survival and growth of Virginia pine planted on acid surface–mine spoil. In *Ecology and Reclamation of Devastated Land*, ed. R. J. Hutnik & G. Davis, pp. 493–504. New York: Gordon & Breach.

Russell, E. W. (1973). *Soil Conditions and Plant Growth*, 10th edn. London: Longman, Green & Co.

Woods, B. (1984). Ants disperse seed of herb species in a restored maple forest (Wisconsin). *Restoration & Management Notes*, **2**, 29–30.

*Eugene B. Welch*
Department of Civil Engineering, University of Washington

*G. Dennis Cooke*
Department of Biological Sciences, Kent State University

# 8 | Lakes

The science and art of restoring ecologically degraded freshwater lakes is a relatively young one. Although projects have been numerous in Europe, a systematic attempt to bring ecological concepts and techniques to bear on the problems of degraded lakes did not begin in earnest in the United States until after 1976, when the Environmental Protection Agency began providing funding for lake restoration. Further encouragement came in 1980, when the EPA established the Clean Lakes Program; several states have since begun their own programs. The result has been a dramatic increase in lake restoration activity, with well over a hundred projects initiated during the the last 10 years (Fig. 8.1). Many of these efforts have been successful, others have been only partially successful, and some have failed completely.

Whatever the outcome from a purely practical point of view, however, this work has proved extremely valuable from a more fundamental point of view, because it has provided unique opportunities for testing ideas about the ecology of lakes. In some cases the results have reinforced existing ideas about the ecology of lakes, but in others they have revealed limitations or weaknesses in those ideas and have stimulated further thinking and research. This in turn has often led to refinements in restoration techniques.

Examples are numerous and pertain to a wide range of basic issues in limnology, including the role of macrophytes in the nutrient economy, the mechanisms of interaction between macrophytes and algae and the role of fish in community structure. Restoration efforts have also led to a major revision of ideas about eutrophication, and overall have brought about fundamental changes in thinking about lake ecology.

While not all limnological research carried out during the last 20 years has been directly related to lake restoration, the challenge of restoring lakes has been a driving force, both from a practical and from a purely intellectual point of view. Indeed it is clear that ultimately the interests of the lake manager are the same as those of the limnologist: both want to be able to manipulate the system with confidence. In this chapter we will explore some examples of how various attempts to restore – or manage – lakes have led to clearer insights into lake ecology.

First, a comment on exactly what is meant by the term restoration when it is applied to lakes. In general restoration refers to any active attempt to

*Fig. 8.1.* Large-scale harvesting, traditionally carried out in an attempt to deal with the problem of dense macrophyte growth in the margins of shallow, eutrophic lakes, actually raises a series of questions about the role these plants play in the ecology of lakes. Thus, for example, technical questions about the best time to harvest interlock with more "basic" questions about the timing of nutrient release by the plants.

return an ecosystem to an earlier condition following degradation resulting from any kind of disturbance. In the case of lakes, however, the most common form of disturbance is the addition of materials that lead to increased productivity and various associated changes in the community. These typically include increases in algal populations, decreases in water transparency, a shift in algal populations from dominance by green algae and diatoms to dominance by blue-greens, a reduction in dissolved oxygen, various changes in fish populations, and ultimately increases in macrophyte populations – all of which tend to be regarded as a decline in lake quality. While lake restoration may address other problems, such as changes in the size or shape of the lake basin, or the presence of exotic plants or animals, lake restoration efforts have tended to concentrate on the problem of accelerated eutrophication and the development of techniques for reversing it and its consequences.

### Bottom-up restoration techniques

Reversing the process of nutrient enrichment of lakes provides an excellent example of the bottom-up approach to lake management. Early observations that the addition of nutrients (especially nitrogen and phosphorus) leads to increased nutrient concentrations in the lake (Hasler, 1947; M. W. Smith, 1969), which in turn were associated with increased algal biomass (Sakamoto, 1966), fostered attempts to reduce productivity in lakes by reducing nutrient input. These experiments have led in turn to important insights into some of the mechanisms controlling productivity and nutrient availability.

Such manipulations to restore lakes typically have been accomplished by diverting nutrient-laden wastewater or stormwater runoff away from the lake and into receiving waters where plant growth is not limited by the nutrients involved (Edmondson, 1970; Ahlgren, 1978). In others, researchers have devised ways to flush the lake with a large volume of nutrient-poor water (Oglesby, 1969; Welch & Patmont, 1980). Both procedures have contributed to a better understanding of how lakes use externally supplied nutrients, and also to useful insights into the relationships between nutrient supplies and the growth of algae and macrophytes. These results, which now enable scientists and managers to manipulate lakes in a far more discriminating manner than before, are summarized below.

*Deep lakes*

*Productivity*

During the early years of intensive lake restoration efforts, the predominant issue was the control of eutrophication in relatively deep lakes with teacup-shaped basins, many of which flush rather rapidly, on the order of one-third or more per year. Lakes of this nature studied following reductions in nutrient input include Lakes Washington, Sammamish and Shagawa in the United States; Lakes Norrviken and Malaren in Sweden; and Lake Zurich in Switzerland. These lakes tended to be viewed as "algae bowls", or flasks containing algae and nutrients. Thus, reducing nutrient input was expected to reduce the nutrient content, and subsequently the algal biomass in the lake. In some instances this is exactly what happened. Lake Washington is the best example. Treated sewage effluent from the city of Seattle was diverted from Lake Washington to Puget Sound during 1963–7. The decline of the phosphorus concentration in Lake Washington following diversion fitted the model Vollenweider (1976) had proposed almost perfectly. That model was based on a simple balance among phosphorus input, output, and loss to the sediments. Furthermore, results of the Lake Washington diversion experiment helped demonstrate that phosphorus, rather than carbon or nitrogen, was the key nutrient in eutrophication, since changes in algal biomass in the lake following diversion declined in direct proportion to phosphorus concentrations and did not parallel changes in either nitrogen or carbon dioxide concentration (Edmondson, 1970; 1972). This was of special interest because both laboratory research by Kerr, Paris & Brochway (1970) and theoretical work by Kuentzel (1969) had suggested a close link between carbon supply and eutrophication. The restoration experiment in Lake Washington and other experiments with whole lakes (Schindler, 1974) made it clear, however, that this is not typically the case and that, at least in some lakes, production is clearly limited by phosphorus entering the lake from outside.

Interestingly, however, experiments on other lakes soon made it clear that reduction in phosphorus input does not invariably lead to a reduction in phosphorus concentrations in a lake. When the EPA used advanced wastewater treatment to reduce the amount of phosphorus entering Lake Shagawa by 85% in 1974, the results were quite unlike those obtained a few years earlier at Lake Washington. Although the average annual phosphorus concentration in the water column did decline significantly, the concentration during the *summer* decreased only slightly, and the amount of algae in the lake remained correspondingly high. A careful analysis of the phosphorus budget and the seasonal dynamics of phosphorus in the lake revealed that large amounts of phosphorus were being released from sediments when the hypolimnion became anaerobic. Moreover, phosphorus released into the hypolimnion became available for algal growth in the lighted epilimnion through storm-induced mixing (Larsen *et al.*, 1979; Larsen, Schultz & Malueg, 1981).

Although it had been appreciated for fifty years that the reduction of iron caused by anoxic conditions can result in the release of phosphorus from sediments (Mortimer, 1941; 1942), the relative ecological importance of this process had not been recognised. The restoration efforts at Lake Shagawa, however, have made it clear that under certain conditions this "internal loading" of phosphorus may be great enough to reduce substantially the response of a lake to decreased *external* loading, and that the internally loaded phosphorus can be utilized by algae even under stratified conditions. An internal loading term is now incorporated into phosphorus budget models, making them more accurate in predicting whole–lake phosphorus concentrations in lakes like Shagawa, at least over the short run (Larsen *et al.*, 1979; Nürenberg, 1984).

It is not yet clear, however, how the internal supply of phosphorus stored in sediments will respond to decreases in external loading in the long term. Ahlgren (1977), for example, found that the release of phosphorus from anaerobic sediments in Lake Norrviken declined over a period of several years following diversion of wastewater and decreased external loading. A similar phenomenon was apparently responsible for a delay of at least seven years in the improvement of water quality in Lake Sammamish following diversion (Welch *et al.*, 1986). In general, the dynamics of the internal phosphorus supply and its release into the water column are still poorly understood, and further study of this phenomenon can be expected to lead to significant improvements in mass balance models for phosphorus in certain lakes, and consequently more cost-effective restorative manipulation.

Of prime importance here, however, is the way in which techniques for reducing the release of phosphorus from the sediment reservoir have provided a way of characterizing this reservoir, the factors that influence its behavior, and the ways in which these vary from lake to lake.

Until recently, for example, conventional limnological wisdom supported the view that the largest source of internal nutrients in lakes is the anaerobic hypolimnetic sediment. Indeed this is the basis for the treatment with aluminum sulfate salts (alum), widely used in lake restoration work to reduce phosphorus release from sediments. In this procedure, an insoluble floc of aluminium hydroxide settles over the sediment, and phosphorus is bound tightly to it, regardless of redox potential. This procedure has proved extremely effective in controlling phosphorus release and in bringing about improved trophic state in most cases (Cooke et al., 1986). The effectiveness of the procedure in reducing phosphorus concentrations in the water column does vary from lake to lake, however; and experience with this restoration technique has made it clear that, while the anaerobic sediments are a major source of internal phosphorus loading, they are not the only source. Aerobic sediments in shallower water can also release phosphorus at a high rate. An alum treatment of the hypolimnetic sediments of West Twin Lake in Ohio, for example, stopped the release of phosphorus from the anaerobic sediment almost completely, but a mass balance analysis revealed that substantial internal loading remained (Cooke et al., 1982). This led to a search for other sources of phosphorus and the identification of senescing macrophyte vegetation and aerobic littoral sediments as possibilities. It is now clear that macrophytes often do play a significant role in moving nutrients from sediments into the water (see p. 118). But numerous recent studies have also shown that considerable amounts of phosphorus are released from sediments into overlying water that is aerobic, especially at the high temperatures and high pH that often occur in littoral zones, or even throughout entire shallow lakes during the summer (Kamp-Nielsen, 1975; Jacoby et al., 1982; Bostrom, Jansson & Forsberg, 1982). It has also become clear that the sediment–water interface may become temporarily anaerobic in polymictic, unstratified lakes under certain conditions. Such phenomena have been inadequately described, however, and carefully designed restoration efforts provide a valuable approach to this task.

Another source of nutrient loading in certain lakes is groundwater percolating upward through littoral sediments. The importance of this is coming to be recognized both as a result of efforts to restore lakes and also as a result of research designed specifically to identify nutrient sources in lakes (Cooke et al., 1982, Prentki et al., 1979). Presumably this water picks up nutrients as it passes through the littoral sediments, and perhaps then transports them into the open-water area. At present we know so little about groundwater and its measurement that limnologists usually estimate the groundwater input into a lake by difference in the hydrologic equation, or else ignore it entirely. LaBaugh & Winter (1984) have drawn attention to some of the errors associated with this approach, however, and we add to their caution the suggestion that groundwater may be an important unknown quantity in the development of nutrient budgets for lakes, not only because of errors in flow estimates, but also because this water may become enriched with both organic and inorganic substances as it passes through the littoral zone.

In any event, these findings suggest that aluminum salts might well be added to the littoral as well as to hypolimnetic sediments in attempts to reduce internal nutrient loading. Several experiments suggest this is indeed the case (Funk & Gibbons, 1979; Jacoby, Welch & Michaud, 1983). Application of alum to littoral habitats or whole lakes, however, poses important questions about aluminum toxicity to lake organisms, a very poorly understood subject.

While the work described above has made it clear that internal sources of phosphorus can play a major role in the maintenance of phosphorus concentrations in a lake, it is also important to understand how phosphorus released from sediments in this way is distributed in the water column. It was commonly believed prior to lake restoration research that phosphorus remains in the anaerobic hypolimnion, inaccessible to algae in the lighted epilimnion, and thus has little influence on algal abundance. In fact, however, Stauffer & Lee (1973) and Larsen, Schultz & Malueg (1981) have shown that considerable phosphorus from the hypolimnion does reach the epilimnion as a result of thermocline erosion during summer storms.

### Community composition

While work such as that described above has led to marked improvements in our ability to predict changes in algal biomass resulting from certain events or manipulations, less progress has been made toward the development of methods for predicting changes in the species composition of the algal community. Nevertheless, prediction of species composition is important in lake management because plankton communities dominated by green algae and diatoms are generally preferred to those dominated by blue-green algae, which form surface scums and may taint the flavor of both fish and drinking water. In addition, prediction of species composition, even in a crude, general way, has long been a goal of lake ecology. This, however, has proved to be extremely difficult. Computer models have so far been of limited help, because the causes for succession are varied and poorly understood. V. H. Smith (1983) suggested that there may be an inverse relationship between the volume-percentage of blue-greens in the algal plankton and the ratio of total nitrogen to total phosphorus in the water column. In other words, a high nitrogen/ phosphorus ratio leads to a low proportion of blue-greens. At N/P ratios greater than 29, blue-green algae are significant in the plankton. The model has little predictive capacity beyond suggesting this critical limit. However, V. H. Smith (1985) has more recently developed a regression model that predicts the concentration of blue-greens on the basis of total nitrogen and total phosphorus content. This effort is extremely important to lake restoration because, in many cases where nutrient inputs cannot be controlled, the only hope may be in manipulating the species composition away from those that cause problems, perhaps through the addition of nitrogen-rich water.

The difficulty of this problem suggests that a combination of factors may influence the relative abundance of green, diatom, and blue-green algae. There is abundant evidence, for example, that a combination of low carbon dioxide concentrations and high pH favors growth of blue-green over green algae, though the reasons for this are not clear. Both King (1972) and Shapiro (1973) stressed the ability of blue-greens to utilize $CO_2$ more efficiently than greens at low $CO_2$ concentrations. It is also true that blue-greens become buoyant and rise to the surface when $CO_2$ concentrations are low and pH high (Paerl & Ustach, 1982), and it is possible that this may account for their competitive advantage under these conditions, since both light and $CO_2$ are more abundant near the surface (Reynolds & Walsby, 1975).

The role of these two mechanisms in controlling the balance between blue-green and green algae is not yet clear. What is clear, however, to some extent as a result of restoration efforts and creation of artificial communities in the laboratory, is that the $CO_2$-pH mechanism is at least partly responsible for the effectiveness of artificial circulation as a technique for shifting the algal community in a lake from one dominated by blue-greens to one dominated by diatoms and green algae. Artificial circulation or destratification was first employed simply to aerate lakes with anoxic hypolimnia in order to increase oxygen for fish. Later Lorenzen & Mitchell (1975) proposed that circulation influences algal biomass by increasing the mixing depth and limiting the light available for photosynthesis. Shapiro, LaMarra & Lynch (1975) and Shapiro (1979a) suggested that mixing raises the carbon dioxide concentration and lowers the pH in the epilimnion, creating conditions under which green algae have the advantage.

Actual experience with this technique, however, has revealed that the effects of circulation on community composition are complex, and that carefully controlled and monitored experiments on whole lakes will be necessary to isolate the factors involved and to determine their relative importance under various conditions. Indeed, until this is done the value of the circulation technique in lake management will be severely limited.

Recently Shapiro *et al.* (1982) and Reynolds, Wisemond & Clarke (1984) have made progress in this direction by carrying out mixing experiments in enclosures *in situ*. The goal of this work is to identify the mechanisms involved in the control of species succession by water column mixing, and to develop prescriptions for the magnitude and frequency of mixing needed to achieve control under various conditions. While artificial circulation is a widely used lake management technique, experiments like these can be expected to increase greatly our understanding of how it works, and so the confidence with which it can be used.

It is worth noting here that while most efforts to reduce algal biomass or to manipulate the composition of algal communities have involved attempts to reduce nutrient supplies or control their distribution, some thought has been given to controlling algae by *adding* nutrients. Nitrate could be added to lakes during the spring diatom bloom, for example, to strip phosphorus from the water column more effectively. This would test the idea that the growth of blue-green algae during the summer depends on residual nutrients remaining in the epilimnion, following nutrient removal by diatoms in the spring.

### Shallow, weed-choked lakes

In general, restoration efforts have played an important role in the development of limnologists' perceptions about the critical differences between lakes, the key processes in lake ecology, and the factors that influence them. Deep, teacup-shaped lakes have been the traditional, textbook model from which many limnologists have learned their trade. However, shallow lakes with extensive, often weed-choked, littoral zones are by far the more common type. Since the littoral zone typically plays a much larger role in the ecology of these lakes than in that of deeper lakes, models based on work with deeper lakes have turned out to be entirely inadequate to account for the behavior of shallow lakes, and efforts to develop more comprehensive models capable of dealing with the more complex behavior of the shallower lakes have recently led to profound changes in our concept of eutrophication. The role restoration has played in this development is complex, and has perhaps not been clearly acknowledged. In our view, however, it has been significant.

One of the early results of lake management efforts, for example, was the finding that reducing nutrient input into a lake does not necessarily lead to a decrease in macrophyte growth. In fact, if it results in an increase in water transparency, it might even cause weed beds to expand, strongly suggesting that their growth is limited by light rather than nutrients. Interestingly, while lake users have long known that accumulations of silt or organic matter in a lake may lead to an expansion of the littoral zone and an increase in macrophyte production, until recently these materials were not included in definitions of the eutrophication process. Obviously, nutrient diversion will have little or no influence on siltation and the accumulation of organic matter. These observations helped to draw attention to the importance of factors other than inorganic nutrients in the ecology of lakes.

At the same time it is important to recognize that much of the research leading to the new ideas has been carried out by scientists interested primarily in basic ecological research rather than in the "applied" problem of lake restoration. The first ecologist to emphasize clearly that most eutrophic lakes are small and shallow, and that many are dominated by the littoral

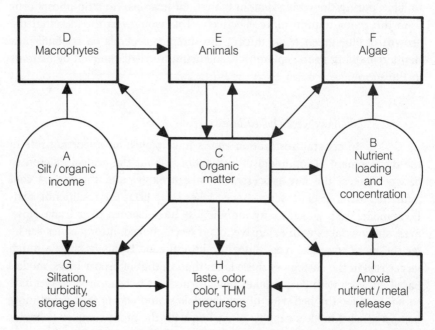

*Fig. 8.2.*   A complex model of the eutrophication process has emerged in recent years, partly as a result of restoration efforts on shallow lakes with extensive littoral zones. These models include inorganic nutrients, but also take account of other factors, notably inputs of silt and organic matter, and positive feedback, which often play a major role in the ecology of smaller lakes. There is evidence that models of this sort are both more accurate and more comprehensive than the older, simpler models, but their validation and refinement will depend on further research, including research involving experimental restoration (Cooke *et al.*, 1986).

zone and the metabolism of detritus produced there was Wetzel, whose research was carried out specifically to characterize the carbon economy of lakes (Wetzel, 1975; Rich & Wetzel, 1978; Wetzel, 1983). Clearly, however, the thinking represented by the conceptual model in Figure 8.2 was based on both "basic" research and experience with lake restoration and management. Reflecting experience with shallow *as well as* deep lakes, this model is both more complex and more comprehensive than the older model of Figure 8.3. Significantly, it includes nutrients, but also emphasizes the role of the silt and organic matter, implying a more complex idea of the eutrophication process, and also suggesting several alternatives besides nutrient diversion a lake manager might use to bring about an improvement in a lake's trophic state.

In fact, neither this model nor the emphasis on silt and organic matter it represents has been generally accepted by limnologists. Of five principle textbooks in limnology (Golterman, 1975; Reid & Wood, 1976; Cole, 1983; Goldman & Horne, 1983; Wetzel, 1983), for example, only Wetzel devotes significant space to the littoral zone community, and two of the authors omit any mention of its ecology. One result of this is that relatively little is known about the littoral community and its plants and animals.

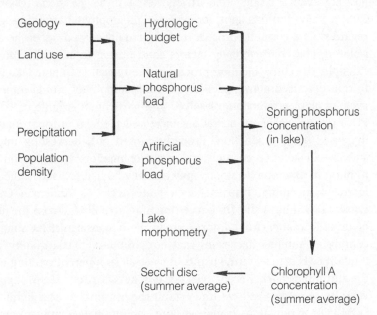

*Fig. 8.3.* Earlier models of lake eutrophication, based largely on experience with deep, teacup-shaped lakes, emphasized nutrient budgets, often to the exclusion of other factors (from Dillon & Rigler, 1975).

Of course ultimately the success of restoration and management efforts based on the model will constitute the "acid test" of its validity. Restoration also offers opportunities for testing ideas about the exact ways in which organic matter influences eutrophication. Already there is evidence that its role is complex. Francko & Wetzel (1981a, b), for example, have shown that cyclic adenosine monophosphate, which may be released from decaying organic matter, stimulates the growth of algae under some conditions. There is also evidence that decaying organic matter releases inorganic nutrients (Heath & Cooke, 1975), lowers oxygen concentrations (which in turn may result in the release of nutrients from sediments) (Bostrom, Jansson & Forsberg 1982), and also increases the shallow, colonizable area in a lake (Carpenter, 1981; 1983). Numerous questions remain, however, about the exact role of these various processes, about their relative importance in lakes of various kinds and about the factors that influence them. There are also important questions about the fate of the organic matter produced. What fraction is deposited on littoral sediments? What fraction accumulates in deeper water? Does the epiphyte community on these plants play an important role in intercepting incoming nutrients?

Well-designed restoration experiments are likely to play an important, possibly even a crucial role in answering these questions (Cooke et al., 1986). Viewed in this light, for example, macrophyte harvesting, generally regarded as a cosmetic procedure, has actually turned out to be a way of isolating and characterizing factors contributing to eutrophication. If, for example, the eutrophication process in the typical eutrophic lake is similar to that described above, then removing macrophytes would interrupt the cycle of plant production, dissolved and particulate organic carbon (DOC-POC) release, sediment accretion, increased internal nutrient loading and increased plant production. Properly carried out, harvesting might help stabilize a lake at its current level of eutrophication. The question is when to harvest. Macrophyte stands may actually have a protective effect if they intercept incoming, nutrient-laden waters. Or, as Anderson, Carlson & Cooke (1985) have shown, harvesting may actually increase the discharge of organic matter from littoral areas. Both considerations suggest that continuous summer harvesting may be detrimental to lake quality, and that it might be best to leave the plants in place all summer, harvesting following the autumn dieback. Recent evidence (Cooke & Carlson, 1986) supports this view. In general, until we understand the seasonality and significance of DOM-POM production, transport, and deposition we cannot know when harvesting would be most effective in protecting the lake. Restoration ecology experiments involving harvesting with appropriate controls, however, ought to be an effective way of answering these questions, and the answers should lead to more effective control techniques.

In fact, restoration experiments of this kind have already led to new insights into the dynamics of lake communities. Virtually complete removal of nuisance macrophytes, either by mechanical harvesting or by grass carp (*Ctenopharyngodon idella*), has been followed by algal blooms (Cooke & Carlson 1986). These blooms have usually been blamed on the technique (e.g., fish excretion of nutrients, sediment entrainment by the harvester, etc.). However, there may be another explanation of far greater significance to restoration ecology. There have been suggestions in the literature for years (Hasler & Jones, 1949) that there is an antagonistic interaction between phytoplankton and macrophytes, so that when macrophytes are removed some factor, which interferes with algal growth, is removed as well and an algal bloom follows. This idea has received little attention in limnology, partly because of the difficulty of testing it experimentally (see Ch. 23). However, as concern about eutrophication has increased, and as procedures for controlling macrophytes in lakes have been developed, the question of this antagonism has been revived. Shireman *et al.* (1983) and Canfield *et al.* (1984) suggest that macrophytes may inhibit phytoplankton in the following ways: through shading; through effective competition for nutrients by epiphytes on macrophyte leaves and filamentous algae associated with the littoral; and through reduced water turbulence, which reduces nutrient transport and thus increases sedimentation of plankton cells. In addition, Wium-Anderson *et al.* (1982) have isolated two allelopathic substances from *Chara* that inhibit phytoplankton metabolism at low concentrations, and van Wiersen & Prins (1985) have demonstrated an allelopathic effect of *Anabaena*, a blue-green alga, on the macrophyte *Zannichellia peltata*. The problem of the exact nature of the hypothesized "antagonism" between phytoplankton and macrophytes is one that deserves further attention, and might well be explored through restoration experiments. Such experiments, designed to identify the factors involved and determine their ecological significance, might well lead not only to important insights into lake ecology, but also to elegant techniques for controlling these plants biologically without the use of chemicals or machinery.

### Top-down techniques

The techniques described so far approach the control of lake productivity from the bottom up, through limitation of available nutrients or light. Of increasing interest, however, are techniques, mostly biological in nature, for exerting control from the "top down," through manipulations that influence organisms, principally fish, near the top of the trophic pyramid. Top-down manipulations have helped identify important relationships between elements of lake communities, and have also led to the development of new ways of manipulating these systems. A new term, "biomanipulation" (Shapiro, 1979a), has been coined to describe these procedures.

Lake restoration studies have played a leading role in the discovery of some of the roles fish play in community dynamics and lake ecosystem function – an issue limnologists, apart from some studies of sport and commercial fish productivity, have traditionally ignored. The following paragraphs describe some of the new ideas about fish that have been discovered either as a direct result of lake restoration efforts, or as a result of studies carried out in connection with them.

The role of certain exotic species, such as the grass carp, in reducing macrophyte biomass directly by grazing has already been mentioned. Other research suggests that fish can influence lake communities by redirecting the movement of nutrients within them and by selective predation, which can influence both the structure and the biomass of the plant community.

### Nutrient budgets

What is the role of fish in the nutrient budget of lakes? While only a few preliminary observations and experiments have been carried out to elucidate this phenomenon, there is evidence that some fish do play significant roles in the internal release of nutrients to the water column, and that others store or conserve nutrients. Using enclosures and microcosms, LaMarra (1975) and Keen & Gagliardi (1981) have shown that the common carp (*Cyprinus carpio*) and the brown bullhead (*Ictalurus nebulosus*), both littoral browsers, excrete large amounts of soluble phosphorus following consumption and digestion of particulate organic matter. LaMarra's measurements suggest that carp alone could introduce phosphorus into the water column at rates similar to the external loading rate associated with the eutrophic state. Similarly, Shapiro *et al.* (1982) found that phosphorus concentrations declined in lakes following the elimination of a large population of carp by winterkill. These preliminary observations suggest yet another source of phosphorus in lake waters, and further illustrate the possible significance of the littoral zone in lake metabolism. In practical terms, they suggest that fish elimination alone could bring about lake improvement in some situations even without expensive diversion of nutrient income. Other evidence suggests that fish may be reserves or sinks for nutrients, adding phosphorus to the hypolimnion through defecation or death (Kitchell, Koonce & Tennis, 1975; Canfield, Maceina & Shireman, 1983), or adding it to the water column directly through decay. Further testing of these ideas through restoration experiments in whole lakes is needed.

Fish may also be involved in the cycling of iron, which in turn may influence the species composition of the algal community. Experiments carried out by Brabrand *et al.* (1984) suggest that some fish are littoral feeders during part of the day and may transport iron to the epilimnion, where it might be expected to increase the growth of algae under conditions where iron is limiting. This work also suggests, however, that fish mucous may chelate iron, tending to remove it from the water column. Since some blue-green algae can also chelate iron, it is plausible that the presence of fish may actually reduce iron availability, possibly favoring iron-chelating blue-green species over other algae, particularly the more desirable green algae, that do not chelate iron. Experiments, possibly involving restoration efforts in whole lakes, may be required to determine which of these effects is ecologically significant under various conditions.

### Selective predation

In addition to their complex and as yet poorly understood role in nutrient cycling, there is also evidence that fish play an important role in shaping the plankton community through selective grazing. Planktivorous fish tend to prefer large-bodied prey species (some species of *Daphnia*) to small-bodied zooplankters (such as *Bosmina*). Consequently, intense planktivory by fish can produce a zooplankton community dominated by small-bodied species (Brooks & Dodson, 1965). Recently, this idea has been tested by experiments involving both enclosures and whole lakes (Stensen *et al.*, 1978; Lynch & Shapiro, 1981, 1982; Shapiro & Wright, 1984). In these studies, populations of large-bodied zooplankton increased, following removal of fish from enclosures, or from lakes by winterkill or rotenone. Similarly, Benndorf *et al.* (1984) have improved lake transparency by adding piscivorous fish, which apparently reduced the density of planktivorous fish, permitting the proliferation of large-bodied herbivorous zooplankton. Indeed, it turns out that under certain conditions algal blooms can be controlled by zooplankton grazing alone, even when nutrient levels are high. In general, these experiments suggest that planktivory may play a major role in the development of plankton communities, which may in turn partly explain why algal biomass, measured by chlorophyll concentrations, is often not closely correlated with nutrient concentrations in lakes (Shapiro, 1979a,b).

The above findings, of course, do suggest ways of working from the top down to reduce the symptoms of eutrophication in a lake. As noted above, one approach is simply to remove fish, thus eliminating one pathway by which nutrients reach the water column. This approach, however, is obviously unacceptable if a lake is to be maintained for fishing. A similar result might be achieved without eliminating fish if zooplankton could be protected from grazing to some extent. Shapiro *et al.* (1982) have suggested that the hypolimnion might provide a refuge for zooplankton, if it were aerated; Timms & Muss (1984) suggested that artificial structures in the littoral zone might serve the same purpose. Clearly this is an area of lake restoration research that is rich in opportunities for experiments that may produce new knowledge about lakes.

In summary, although it is not altogether certain that selective planktivory really is responsible for the fact that lakes with algal blooms often have small-bodied zooplankton (Webster & Peters, 1978; Gliwicz & Siedlar, 1980; Porter, 1981), this phenomenon is now well enough understood to provide the basis for management procedures under favorable conditions.

### Forests and lakes: restoration and ecological theory

We would like to conclude this discussion of the basic ecological value of lake restoration efforts by describing how such efforts have led to recognition of a fundamental similarity between two apparently quite different ecosystems, forests and the littoral zones of eutrophic lakes. Both of these systems have a three-tiered, detritus-based trophic structure in which fourth-level consumers such as hawks, owls, bass or pike are absent, or present only in small numbers. As a result populations of third-level consumers (insectivorous birds or sunfish) are abundant and maintain second level consumers (herbivores such as insects or plant eating fish) at a low level, so that biomass accumulates at the producer level.

In this sense, then, the littoral zone more closely resembles the forest ecosystem than the adjacent, grazer-based pelagic zone, where fourth-level consumers keep populations of third-level consumers low, so that herbivore populations are large, and a large fraction of primary production is consumed by herbivores (such as *Daphnia*).

In general, the normal course of events in ecological succession seems to be away from the four-level, grazer-based system toward the three-level, detritus-based system (Cooke, 1967; Odum, 1969; 1971), with many feedback loops to stabilize and maintain it. Restoration of lakes, however, typically means reestablishing the grazer-based system. This can be done either by increasing fourth-level predator populations, or perhaps by eliminating the third trophic level entirely. Either approach might result in the even number of trophic levels characteristic of a grazer-based system, although of course the two-level approach is out of the question in a lake that is to be restored to a natural three- or four-layered condition, or if it is simply being managed for fish. In any case, experiences with degradation of lakes and their restoration lend support to this concept of trophic structure. For example, even the plankton community can become a three-level system, producing large algal blooms, when cold-water fisheries are eliminated and populations of predator fish decline, or if littoral refuges for third-level fish are created (F. E. Smith, 1969). In contrast, biomanipulation, by increasing populations of predators, can return the community to a four-level one in which herbivores control biomass (Shapiro, 1979a).

In some cases this may be practical. However, it is likely to be much easier in deep, teacup-shaped lakes than in the more common saucer-shaped lakes. Since shallow lake communities lack large populations of endemic herbivores, their conversion to two-or four-level systems would mean either bringing in exotic herbivores (like grass carp, or cattle), or reshaping the lake basin to eliminate the littoral zone, a form of manipulation that is analogous to using a plow to convert a detritus-based forest or old-field system into a grazer-based agricultural system (see Harper's comments on the relationship between agriculture and ecological research, in Ch. 3). This may change the appearance of the system, and may even retard succession in it. However, since, in contrast with the top-down approach, it does not alter the trophic state of the system, it creates a system under stress. This is not restoration in the usual sense. Of special interest here, however, is the way in which restoration efforts have lent support to generalizations that draw attention to underlying similarities between communities – clearly an important part of the process of developing a science and art of ecological repair.

### References

Ahlgren, I. (1977). Role of sediments in the process of recovery of a eutrophicated lake. In *Interactions Between Sediments and Fresh Water*, ed. H. L. Golterman, pp. 327–7. The Hague: Dr W. Junk Publications.

Ahlgren, I. (1978). Response of Lake Norrviken to reduced nutrient loading. *Verh. Int. Ver. Limnol.*, **20**, 846–50.

Anderson, P. W., Carlson, R. E. & Cooke, G. D. (1985). An input-output study of nutrient loading by macrophytes: Do enclosures reflect reality? *Abstracts, American Society of Limnology and Oceanography.*

Benndorf, J., Kneschke, H., Kossatz, K. & Penz, E. (1984). Manipulation of the pelagic food web by stocking with predacious fishes. *Internationale Revue der gesamten Hydrobiologie*, **69**, 407–28.

Bostrom, B., Jansson, M. & Forsberg, C. (1982). Phosphorus release from lake sediments. *Archiv für Hydrobiologie Beih. Ergebn. Limnol.*, **18**, 5–59.

Brabrand, A., Faafeng, B., Kollquist, T. & Nilssen, J. T. (1984). Can iron defecation from fish influence phytoplankton production and biomass in eutrophic lakes? *Limnology and Oceanography*, **29**, 1330–4.

Brooks, J. L. & Dodson, S. I. (1965). Predation, body size, and composition of plankton. *Science*, **150**, 28–35.

Canfield, D. E. Jr., Maceina, M. J. & Shireman, J. V. (1983). Effects of hydrilla and grass carp on water quality in a Florida lake. *Water Resources Bulletin*, **19**, 773–8.

Canfield, D. E. Jr., Shireman, J. V., Colle, D. E., Holler, W. T., Watkins, C. E. II & Maceina, M. J. (1984). Prediction of chlorophyll *a* in Florida lakes: Importance of macrophytes. *Canadian Journal of Fisheries and Aquatic Sciences.*, **41**, 497–501.

Carpenter, S. R. (1981). Submersed vegetation: an internal factor in lake ecosystem succession. *American Naturalist*, **118**, 372–83.

Carpenter, S. R. (1983). Lake geometry: implications for production and sediment accretion rates. *Journal of Theoretical Biology*, **105**, 273–86.

Cole, G. A. (1983). *Textbook of Limnology*. St Louis: C. V. Mosby.

Cooke, G. D. (1967). The patterns of autotrophic succession in laboratory microcosms. *BioScience*, **17**, 717–21.

Cooke, G. D., Heath, R. T., Kennedy, R. H. & McComas, M. R. (1982). Change in lake trophic state and internal phosphorus release after aluminum sulfate application. *Water Resources Bulletin*, **18**, 699–705.

Cooke, G. D. & Carlson, R. E. (1986). Water quality management in a drinking water reservoir. In *Applied Lake and Watershed Management*. Virginia: Proceedings of North American Lake Management Society Symposium.

Cooke, G. D., Welch, E. B., Peterson, S. A. & Newroth, P. R. (1986). *Lake and Reservoir Restoration*. Boston: Butterworth.

Dillon, P. J. & Rigler, F. H. (1975). A simple method for predicting the capacity of a lake for development based on lake trophic status. *J. Fish. Res. Bd. Can.*, **32**, 1519–31.

Edmondson, W. T. (1970). Phosphorus, nitrogen and algae in Lake Washington after diversion of sewage. *Science*, **169**, 690–1.

Edmondson, W. T. (1972). Nutrients and phytoplankton in Lake Washington. In *Nutrients and Eutrophication: The Limiting Nutrient Controversy*, Spectral Symposium, ed. G. E. Likens vol. 1, pp. 172–93. American Society of Limnology and Oceanography.

Francko, D. A. & Wetzel, R. G. (1981a). Synthesis and release of cyclic adenosine 3′:5′-monophosphate by aquatic macrophytes. *Physiologia Plantarum*, **52**, 33–6.

Francko, D. A. & Wetzel, R. G. (1981b). Dynamics of cellular and extracellular cAMP in *Anaboena flos-aquae* (Cycanophyta): Intrinsic culture variability and correlation with metabolic variables. *Journal of Phycology*, **17**, 129–34.

Funk, W. H. & Gibbons, H. L. (1979). Lake restoration by nutrient inactivation. In *Lake Restoration*, pp. 141–51. EPA 400/5–79–001.

Gliwicz, A. M. & Siedlar, E. (1980). Food size limitation and algae interfering with food collection in *Daphnia*. *Archiv für Hydrobiologie*, **88**, 155–77.

Goldman, C. R. & Horne, A. J. (1983). *Limnology*. New York: Elsevier Scientific.

Golterman, H. L. (1975). *Physiological Limnology*. New York: Elsevier Scientific.

Hasler, A. D. (1947). Eutrophication of lakes by domestic drainage. *Ecology*, **28**, 383–95.

Hasler, A. D. & Jones, E. (1949). Demonstrations of the antagonistic action of large aquatic plants on algae and rotifers. *Ecology*, **30**, 359–64.

Heath, R. T. & Cooke, G. D. (1975). The significance of alkaline phosphatase in a eutrophic lake. *Verhandlungen Internationale Vereinigung Limnologie*. **19**, 959–65.

Jacoby, J. M., Lynch, D. D., Welch, E. B. & Perkins, M. A. (1982). Internal phosphorus loading in a shallow eutrophic lake. *Water Research*, **16**, 911–13.

Jacoby, J. M., Welch, E. B. & Michaud, J. P. (1983). Control of internal phosphorus loading in a shallow lake by drawdown and alum. In *Lake Restoration, Protection and Management*, pp. 112–8. EPA 440/5–83–001.

Kamp-Nielsen, L. (1975). Seasonal variation in sediment-water exchange of nutrient ions in Lake Esrom. *Verhandlungen Internationale Vereinigung Limnologie*, **19**, 1057–65.

Keen, W. H. & Gagliardi, J. (1981). Effect of brown bullheads on release of phosphorus in sediment and water systems. *Progressive Fish-Culturist*, **43**, 183–5.

Kerr, P. C., Paris, D. F. & Brochway, D. L. (1970). The interrelation of carbon and phosphorus in regulating heterotrophic and autotrophic populations in aquatic ecosystems. EPA Rep. 16060 FGS 07/70.

King, D. L. (1972). Carbon limitation in sewage lagoons. In *Nutrients and Eutrophication: The Limiting Nutrient Controversy*, Special Symposium, ed. G. E. Likens, vol. 1, pp. 98–110. American Society of Limnology and Oceanography.

Kitchell, J. F., Koonce, J. F. & Tennis, P. S. (1975). Phosphorus flux in fishes. *Verh. Int. Ver. Limnol.*, **19**, 2478–84.

Kuentzel, L. E. (1969). Bacteria, carbon dioxide and algal blooms. *Journal of the Water Pollution Control Federation*, **41**, 1737–47.

LaBaugh, J. W. & Winter, T. C. (1984). The impact of uncertainties in hydrologic measurement on phosphorus budgets and empirical models in two Colorado reservoirs. *Limnology and Oceanography*, **29**, 322–39.

LaMarra, V. J. Jr (1975). Digestive activities of carp as a major contributor to the nutrient loading of lakes. *Verh. Int. Ver. Limnol.*, **19**, 2461–8.

Larsen, D. P., Van Sickley, J., Malueg, K. W. & Smith, P. D. (1979). The effect of wastewater phosphorus removal on Shagawa Lake, Minnesota: Phosphorus supplies, lake phosphorus and chlorophyll *a*. *Water Research*, **13**, 1259–72.

Larsen, D. P., Schultz, D. W. & Malueg, K. W. (1981). Summer internal phosphorus supplies in Shagawa Lake, Minnesota. *Limnology and Oceanography*, **26**, 740–53.

Lorenzen, M. W. & Mitchell, R. (1975). An evaluation of artificial destratification for control of algal blooms. *Journal of the American Water Works Association*, **67**, 373–6.

Lynch, M. & Shapiro, J. (1981). Predation, enrichment, and phytoplankton community structure. *Limnology and Oceanography*, **26**, 86–102.

Lynch, M. & Shapiro, J. (1982). Manipulations of planktivorous fish-effects on zooplankton and phytoplankton. In *Experiments and Experiences in Biomanipulation*,

128     E. B. Welch and G. D. Cooke

Report No. 19, ed. J. Shapiro *et al.*, pp. 158–89. Minneapolis: University of Minnesota.

Mortimer, C. H. (1941). The exchange of dissolved substances between mud and water in lakes (Parts I and II). *Journal of Ecology*, 29, 280–329.

Mortimer, C. H. (1942). The exchange of dissolved substances between mud and water in lakes (Parts III, IV, summary, references). *Journal of Ecology*, 30, 147–201.

Nürenberg, G. K. (1984). The prediction of internal phosphorus load in lakes with anoxic hypolimnia. *Limnology and Oceanography* 29, 111–25.

Odum, E. P. (1969). The strategy of ecosystem development. *Science*, 164, 262–70.

Odum, E. P. (1971). *Fundamentals of Ecology*, 3rd edn. Philadelphia: Saunders.

Oglesby, R. T. (1969). Effects of controlled nutrient dilution on the eutrophication of a lake. In *Eutrophication: Causes, Consequences and Correctives*, pp. 483–93. Washington DC: National Academy of Sciences.

Paerl, H. W. & Ustach, J. F. (1982). Blue–green algal scums: An explanation for their occurrence during freshwater blooms. *Limnology and Oceanography*, 27, 212–17.

Porter, K. G. (1981). Limits to the control of algal populations by grazing zooplankton. In *Proceedings of Workshop on Algal Management and Control*, Technical Report A–81–7., pp. 121–30. Vicksburg: US Army Corps of Engineers.

Prentki, R. T., Adams, M. S., Carpenter, S. R., Gasith, A., Smith, C. S. & Weiler, P. R. (1979). The role of submerged weedbeds in internal loading and interception of allochthonous materials in Lake Wingra, Winconsin, USA. *Archives of Hydrobiology/Supplement*, 57, 2, 221–50.

Reid, G. K. & Wood, R. D. (1976). *Ecology of Inland Waters and Estuaries*. New York: van Nostrand.

Reynolds, C. S. & Walsby, A. E. (1975). Water blooms, *Biological Reviews*, 50, 437–81.

Reynolds, C. S., Wisemond, S. W. & Clarke, M. J. (1984). Growth– and loss–rate response of phytoplankton to intermittent artificial mixing and their potential application to the control of planktonic algal biomass. *Journal of Applied Ecology*, 21, 11–29.

Rich, P. H. & Wetzel, R. G. (1978). Detritus in the lake ecosystem. *American Naturalist*, 112, 57–71.

Sakamoto, M. (1966). Primary production by phytoplankton community in some Japanese lakes and its dependence on lake depth. *Archiv für Hydrobiologie*, 62, 1–28.

Schindler, D. W. (1974). Eutrophication and recovery in experimental lakes: implications for lake management. *Science*, 184, 897–9.

Shapiro, J. (1973). Blue–green algae: why they become dominant. *Science*, 179, 381–4.

Shapiro, J. (1979a). The need for more biology in lake restoration. In *Lake Restoration*. EPA 440/5–79–001.

Shapiro, J. (1979b). The importance of trophic level interactions to the abundance and species composition of algae in lakes. SIL Workshop on Hypertrophic Ecosystems. *Developmental Hydrobiology*, 2, 105–16.

Shapiro, J., Forsberg, B., LaMarra, V., Lindmark, G., Lynch, M., Smeltzer, E. & Zoto, G. (1982). *Experiments and Experiences in Biomanipulation*, Interim Report No. 19. Mineapolis: University of Minnesota.

Shapiro, J., LaMarra, V. & Lynch, M. (1975). Biomanipulation: An ecosystem approach to lake restoration. In *Water Quality Management Through Biological Control*, ed. P. L. Brezonik & J. L. Fox, pp. 85–96. Gainesville: University of Florida.

Shapiro, J. & Wright, D. I. (1984). Lake restoration by biomanipulation: Round Lake, Minnesota, the first two years. *Freshwater Biology*, **14**, 371–83.

Shireman, J. V., Haller, W. T., Colle, D. E., Watkins, C. E., Durant, D. F. & Canfield, D. E. (1983). Ecological impact of integrated chemical and biological aquatic weed control. EPA 660/3–83–098.

Smith, F. E. (1969). Effects of enrichment in mathematical models. In *Eutrophication: Causes, Consequences and Correctives*, pp. 631–45. Washington DC: National Academy of Sciences.

Smith, M. W. (1969). Changes in environment and biota of a natural lake after fertilization. *J. Fish. Res. Bd. Canda*, **26**, 3101–32.

Smith, V. H. (1983). Low nitrogen to phosphorus ratios favor dominance by blue-green algae in lake phytoplankton. *Science*, **221**, 669–71.

Smith, V. H. (1985). Predictive models for the biomass of blue-green algae in lakes. *Water Resources Bulletin*, **21**, 433–9.

Stauffer, R. E. & Lee, G. F. (1973). The role of thermocline migration in regulating algal bloom. In *Modelling the Eutrophication Process*, ed. D. H. Falkenborg & T. E. Maloney, pp. 73–82. Logan: Utah State University.

Stensen, J. A. E., Bohlen, T., Henrikson, L., Nilsson, B. I., Nyman, H. G., Oscarson, H. G. & Larsson, P. (1978). Effects of fish removal from a small lake. *Verh. Int. Ver. Limnol.*, **20**, 794–801.

Timms, R. M. & Muss, B. (1984). Prevention of growth of potentially dense phytoplankton populations by zooplankton grazing, in the presence of zooplantivorous fish, in a shallow, wetland ecosystem. *Limnology and Oceanography*, **29**, 472–86.

van Wiersen, W. & Prins, T. C. (1985). On the relationship between the growth of algae and aquatic macrophytes in brackish water. *Aquatic Botany*, **21**, 165–79.

Vollenweider, R. A. (1976). Advances in defining critical loading levels for phosphorus in lake eutrophication. *Mem. Ist. Ial. Idrobiol.*, **33**, 53–83.

Webster, K. E. & Peters, R. H. (1978). Some size–dependent inhibitions of larger cladoceran filterers in filamentous suspensions. *Limnology and Oceanography*, **23**, 1238–45.

Welch, E. B. & Patmont, C. R. (1980). Lake restoration by dilution: Moses Lake, Washington. *Water Research*, **14**, 1317–25.

Welch, E. B., Spyridakis, D. E., Shuster, J. I. & Horner, R. R. (1986). Declining lake sediment phosphorus release and oxygen deficit following wastewater diversion. *Journal of the Water Pollution Control Federation*, **58**, 92–6.

Wetzel, R. G. (1975). *Limnology*. Philadelphia: W. B. Saunders.

Wetzel, R. G. (1983). *Limnology*. Philadelphia: W. B. Saunders.

Wium-Anderson, S., Anthoni, U., Christopherson, C. & Hoven, G. (1982). Allelopathic effects on phytoplankton by substances isolated from aquatic macrophytes (Charales). *Oikos*, **39**, 187–90.

# III | Synthetic ecology

...raw reduction is only half the scientific process. The
remainder consists of the reconstruction of complexity by an
expanding synthesis under the control of laws newly
demonstrated by analysis.

E. O. Wilson *On Human Nature*

While the actual assembly of ecological communities in the field
has rarely been regarded as a form of ecological experimentation,
there is a substantial tradition of ecological research involving
the creation of systems in the laboratory specifically in order to
test ideas about them. It is this tradition and its possible relationship to restoration in its more familiar forms that we will
explore in this section. The two chapters in this section represent
a considerable range of ecological issues. The first, Walter Adey's
account of his experiences with the construction of artificial reef-
lagoon ecosystems at the Smithsonian Institution, deals with
both community and ecosystem issues. Mike Gilpin's chapter, in
contrast, is exclusively community-oriented. In addition, Adey's
work, though it turned out to be richly rewarding in a more
fundamental sense, was undertaken in order to create the
system to serve as a subject for research – a product-oriented
approach not unlike the approach that characterized the early
work at the UW Arboretum (see Ch. 1). Gilpin's work, in contrast, produced a product of no value in itself. It was carried out

specifically to test a single hypothesis and represents the only example of such a project included in this book. In either case, however, our interest is not so much on what can be learned from the completed system as from the process of creating it.

*Walter H. Adey*

Marine Systems Laboratory, Smithsonian Institution

# 9 | Marine microcosms

This chapter describes the construction and refinement of an artificial tropical coral reef and lagoon ecosystem. Emphasis is placed on the extensive process of synthesizing such systems through prototype and pilot stages, and on the insights that this work has added to our understanding of wild ecosystems.

We have learned a great deal about the ecology of marine ecosystems as a direct result of efforts to design, engineer, construct and refine such systems in microcosm or mesocosm form. In many cases, the results of learning from system syntheses were surprising, and contradicted prevailing ideas based on mathematical models and/or field work. By its very nature, a microcosm must be assembled piece by piece, beginning with an engineering phase, followed by addition of the community, and finally by population and ecological manipulations, assisted by additional engineering design and construction.

Our experiences in carrying out field work interactive with microcosm syntheses make it clear that microcosms and mesocosms, in addition to being valuable heuristic tools, can also become, in effect, prototype and pilot systems for actual restoration work. Used by the restorationist such systems could provide insights and answers to questions at every step, and could also dramatically improve chances of success.

Basta & Moreau (1982), in their modern treatment *Introduction to Analyzing Natural Systems*, reduce the study of ecosystems to three basic methods: "(1) physical (scale) modeling; (2) conservation of mass and energy approaches; and (3) statistical methods." Other scientists would perhaps use a somewhat different terminology, referring to conservation of mass and energy approaches as mathematical modeling, and to Basta and Moreau's "statistical methods" simply as natural history or field studies. However, the conceptual framework for understanding ecosystems is more or less universally accepted, and it is clear that each of these approaches has both strengths and weaknesses, as well as its critics and practitioners.

The traditional, statistical methods, involving collecting data from "normal" systems, relating variables by regressions and looking for variance from the norm clearly has great value, and in some ways it is indispensable. Its disadvantage is that it often depends on the collection of data over impractically long periods of time. The long-standing argument over dramatic increases in populations of the coral eating starfish (*Acanthaster*) in the Indo-Pacific is an example. Some ecologists say that we are witnessing an unnatural population explosion, probably brought on by human interference, and perhaps by over-collection of its snail predator. The unfortunate truth is that we simply do not know in any time scale beyond a few decades whether the "explosion" is even statistically abnormal.

The conservation of mass and energy methods are standard to modeling science. Many would regard them as the only modern approach. For example, the recent 600-page modeling volume from *The Sea* series includes only mathematically oriented conservation of mass and energy approaches (Goldberg *et al.*, 1977). Energy is required to drive biological systems, and one can certainly argue that energy capture, storage and eventual degradation is the most important characteristic of an ecosystem. However, this does not mean that, ultimately, mapping energy flow through a system provides a predictive understanding of how that ecosystem will behave when perturbed, or even when it is responding normally to some change in the environment or some event within the system itself. Would any presently conceivable energy or material flow models have made it possible to predict the *Acanthaster* population increase or the recent dramatic decline of sea urchin (*Diadema*) populations in Caribbean reefs – or even the consequences of these changes?

Of the third general approach to the study of natural systems, variously called physical or scale modeling, microcosm, microecosystem or mesocosm modeling, living simulation modeling, or, in our context, ecological synthesis, Basta & Moreau (1982) state simply, "this does not mean that physical modeling is inferior to the other two approaches. To the contrary, in some situations it is superior, and in a few situations it may be the only operational approach".

Certainly, as noted in the introduction to this book, this is true of various non-biological systems. Hydraulic engineers (e.g. Hudson, 1979) note that engineers have used physical models that simulate the hydrology of harbors and estuaries in design work for over a century. Today such models number in the hundreds, and have been built and used for virtually every body of water that is subject to heavy human use. Naval architects, with their extensive – and expensive – ship model towing basins, often point out that model towing in ship design preceded naval architecture as an engineering field (Stuntz, 1963).

Physical modeling techniques in engineering were developed and survive today, along with and alongside modern mathematics and computers, because they answer practical questions about very complex systems that cannot be approached in other ways, or that are more conveniently handled by such models. While the physical and mathematical configuring of simple hydraulic systems (pipes, cylinders, trenches) has been developed to some degree of sophistication, and there are computers that can handle the extensive computations generated by such analyses, a real bay or ship hull is so complex physically that a rigorous mathematical modeling of tidal, salinity, or wave effects, or of hull motion, often remains technically impractical. In short, physical modeling, or "synthesis", is especially useful – and may be indispensible – in dealing with complex systems that defy rigorous mathematical analysis, a fact that points directly towards the value of this approach to the study of complex biological systems such as ecological communities and ecosystems. A suggestive analogy is offered by research on the brain, which has profited immensely from the development or synthesis of a physical model – the computer – and its use in research on brain function (Gardner, 1985).

What I would like to describe in this chapter are experiences with microcosm modeling of three marine ecosystems. While the value of this approach has been debated, our own experience, and that of others in the field, makes it clear that working with, and especially trying to construct, physical models of these systems can lead to numerous insights, some quite surprising. I feel that we might not have achieved these without the hands-on experience of actually putting systems together and working with them until they were functioning properly.

This chapter is a brief introduction to the world of marine ecosystem simulation. Above all, it is intended to demonstrate the heuristic "mind and hands" value of actual system synthesis in the hope of stimulating more general interest in this approach to ecological research.

### Marine microcosms

In this chapter I will describe in some detail my experiences in the construction and refinement of several marine microcosms, including several Caribbean coral reef microcosms of increasing complexity, developed in my laboratory over nearly ten years of interactive research and synthesis. This discussion will stress the heuristic value of the actual construction of these systems. It will especially emphasize instances in which the results were unexpected and provided insights into the functioning of the wild ecosystem. An extensive review of marine microcosm practice as related to ecological research is provided by Pilson & Nixon (1980). Here, in order to demonstrate the broad applicability of this approach, I have chosen projects representing work with several widely different ecosystem types.

There are two basic approaches to developing marine microcosms: aquaria (of a wide variety of shapes, sizes and materials); and on-site enclosures. Either type can be closed or open, with controlled access to a wild system. In an aquarium microcosm, one provides all of the required physical components in an appropriate sequence: light energy (artificial or natural); physical heat balance (temperature control); hydraulic energy (wave, current and tidal control); salt balance (evaporation control); geological base (substrate, if benthic); seawater (of appropriate density); the community of organisms (in roughly the appropriate numbers of individuals); and finally, and perhaps most critical and difficult, the biological, chemical and physical effects of adjacent ecosystems. In general, bringing together a community of organisms under appropriate physical conditions is not difficult in these systems, though it often requires considerable engineering ingenuity. The problems arise largely in establishing the interactive effects of adjacent ecosystems and in scaling. These two problems are closely related, since scale typically involves establishing trophic and competitive balances when some organisms or their territories are simply too large for the system being constructed.

Perez et al. (1977) discuss in some depth the problem of scaling in microcosm simulation. It is worth noting here, however, that construction of these systems offers many opportunities for testing ideas about the importance of scaling in ecosystems, as discussed by Allen & Hoekstra in Chapter 19. Ultimately the problem of scale arises as a result of economic and practical considerations. Since microcosms are never as large as the systems they represent, it is necessary to simulate or account for the effects of larger size. This forces us to consider very carefully the importance of space- and time-related factors in the systems being modeled. This can also lead to unexpected and instructive complications. A higher predator, such as a barracuda, in a coral reef ecosystem, for example, needs a large territory. If such a rapacious fish is included in a small model system, it will have to be fed with imported small fish to prevent overcropping the "support community". This in turn introduces excess nutrients into the system. But these can be more or less controllably removed, as discussed on page 143.

The issue of interaction between systems is perhaps the same – if we can manufacture one ecosystem, we can also construct the adjacent system to provide the appropriate interactions. Then, however, the exact nature and magnitude of the interactions become crucial, and the question inevitably becomes, where do we stop, short of a complete biosphere, in our additions of adjacent ecosystems? As in the matter of scale, the problems are often solved not by a literal reproduction of the adjacent ecosystem, but by learning to reproduce the functional effects of that system.

This can be highly instructive: reefs, for example, generally receive their overlying waters from the open ocean and release waste water downstream to lagoons. Most lagoon waters are ejected through reef passes and undergo extensive mixing with ocean water before again impinging on a reef. The effect of this is to provide very low nutrient content in the overlying water, even though reef biomass is very high. Thus, in our research, an algal scrubber was devised to simulate the effects of the ocean in maintaining water quality in a reef community with a naturally "excessive" rate of metabolism. This in turn led to insights into the role of the ocean in reef ecology, and to the understanding that the large biomass, respiration and nutrient requirements of reefs can be maintained only with support from the adjacent ocean waters. In general, therefore, placing glass walls around a piece of a wild reef would quickly be disastrous to that community.

It should be pointed out that there is little or no relationship between the traditional marine aquarium and a microcosm designed to simulate more or less accurately ecosystem function. The traditional aquarium provides only the most crucial needs (basically food, toxic waste removal, and oxygen subsistence levels) for fish, and perhaps a few larger invertebrates. The inevitable disease problems of stressed animals in a highly artificial environment are usually treated by chemicals such as ozone or copper which leave the aquarium substerile, except for the organisms for which the tank is maintained. As one might guess, aquaria are far less stable than well-designed microcosms. This is a profound difference that is not necessarily simply the result of the greater biotic complexity of the microcosm. In fact, extremely simple biological systems have been maintained in small microcosms for extended periods.

Even though the end result is the same, enclosures are conceptually quite different from aquaria. By their very nature, they include all of the appropriate factors. The primary question is, how have natural processes been blocked or altered by the addition of the enclosure? Thus, at least until enclosure design becomes much more sophisticated, enclosure modeling will probably be most useful for investigating relatively short-term phenomena.

### The tropical coral reef and lagoon microcosm

Coral reef ecosystems, perhaps together with rainforests, have been described as "... form(ing) a class of complex ecosystems which are neither qualitative or even quantitative extensions of simpler communities. They are systems of a higher order..." (Bradbury, 1977). Whether one agrees with this statement or not, reef communities are certainly very complex, and lie at one end of the spectrum of complexity for marine ecosystem types. Partly for this reason there have been fewer serious attempts to model wild reefs mathematically than certain other marine systems, and most of these have been undertaken very recently (e.g., Grigg, Polovina & Atkinson, 1984). Attempts to model reefs physically have also been very limited, ranging from microbial mini-microcosms (Kearns & Folsome, 1981) to small aquaria (Henderson, Smith & Evans, 1976).

The magnitude of primary production in coral reefs, the organisms responsible for that production, and also the effects of nutrient control on it have been widely discussed and disputed (see Lewis, 1977 and Adey & Steneck, 1985 for extensive discussion). It has generally been felt that primary production on reefs, as in the open ocean in general, is limited to moderate levels by low nutrient concentrations and the necessity, at least in the absence of an outside source such as upwelling, of recyling all required nutrients within the system.

It was while teaching field courses on the role of algae in reef ecosystems on Saint Croix in the Virgin Islands in 1975 and 1976 that I became aware that the interaction of the physical ocean environment with the smaller algal species (algal turfs) in reef communities was likely to be the key to reef function. Thus, beginning in 1976, I initiated construction of a 240 l "reef" aquarium in Washington DC, beginning with the premise that algal turfs are the cornerstone elements of shallow reef systems. Initially, the effort was largely autecological in emphasis, since my primary interest was to maintain a diverse algal turf assemblage for study. However, algal turfs on reefs are an early successional stage that is normally maintained (as is grassland sometimes) by heavy grazing by a host of herbivores. Thus, it was necessary to add these species. As the complexity of the system increased, it soon became synecological, or ecosystem-simulation oriented. Some success with this early unit led to the development of a series of reef microcosms of increasing complexity, as described below.

In 1983, I described a 7 kl closed microcosm that had been operated successfully for about seven years before being demolished because of space requirements (Adey, 1983). Here I will briefly discuss the smaller prede- cessor systems, and then finally a successor system of 12 kl that has been functioning for about six years (Figs 9.1, 9.2). The 12 kl unit is a good example of both the advantages and the problems associated with both scaling and the interaction of adjacent systems, since it is still relatively small and actually comprises two adjacent ecosystems – the reef itself and the lagoon. The basic parameters of these model ecosystems are described by Adey (1983) and Williams & Adey (1983) and will not be described here. The present list of organisms in these systems includes more than 300 species, but is a minimum figure since several major taxonomic groups (including bacteria, sponges and nematodes) and a few lesser groups have not been analyzed in detail. In my earlier report (Adey, 1983) I concluded that on the basis of its species diversity, this small system simulated 20 $m^2$ rather than the roughly 4 $m^2$ actually present in the model. The present microcosm represents an attempt to scale about 600 $m^2$ of a wild reef and lagoon in the same space, and thus is a 1/150 scale model physically and a 1/30 scale model biologically.

The process of developing these systems provides an excellent example of the heuristic nature of system synthesis. The prototype system was a relatively simple, 240 l natural-light greenhouse unit at the Smithsonian Institution. Temperature- and salinity- controlled, and having current action, this small microcosm functioned reasonably well for such a small

system except during the long, cloudy, low-light periods of autumn and winter. While light levels even during cloudy intervals were considerably higher than most scientists would have predicted to be optimum for the system, on the basis of extensive field studies of plankton and laboratory studies with benthic algae, we found that even during daylight hours photosynthesis in our tank was insufficient to support the heterotrophic community. We later found a similar, though less pronounced, effect during autumn on wild reefs in the Caribbean (Adey & Steneck, 1985).

Having established this, we designed our next (1 kl) tank with a set of mercury vapor lamps on a 12-hour photoperiod to replace the unreliable natural lighting of our area. This allowed us to maintain some of the less sensitive corals (an indication of improved environmental conditions), and led to a considerable increase in the species diversity of the system generally. However, photosynthesis (as measured by oxygen exchange) remained well below respiration on a diel basis. since the light produced by mercury vapor lamps is richer in blue and ultraviolet than earth-surface solar radiation, and since ultraviolet is recognized as a photosynthetic inhibitor, after about eight months we replaced the mercury vapor lamps with metal halide

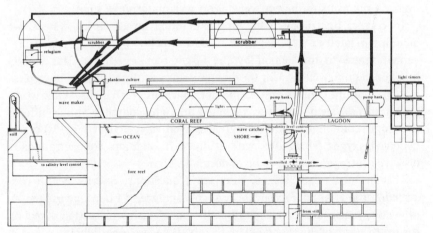

*Fig. 9.1.* Functional layout and photograph of the 12 kl coral reef/ lagoon microcosm currently operating both as a research system and as a public display at the Smithsonian Institution. Like many other eco- system construction projects originally undertaken in order to create systems to serve as subjects for research or for some practical or environmental purpose, this system has demonstrated the fundamental value of the process of construction itself (photograph courtesy of the Smithsonian Institution).

lamps, which produce light that more closely matches the intensity and spectrum of tropical solar radiation. Use of 400-watt (and in deeper sections, 1000-watt) metal halide lamps now provides light intensities of about 1100 uE/m²/s at the reef surface. This is somewhat below the 1200–1800 uE/m²/s intensity typically found in summer on shallow Caribbean reefs, and while photosynthetic rates were similar to those of some wild reefs, they were still well below those of the most productive reefs. A new metal halide lamp with an output about 30 per cent higher than the present lamps will soon be tested in our 12 kl system. This factor will be discussed in more detail below.

Having only a small lagoon adjacent to the reef, but possessing a richer reef community than had been previously tried, and achieving reef-level photosynthetic rates only under intense radiation, our 1 kl system was highly successful and operated for about two years until glass failure terminated the operation. However, the lagoon was subject to heavy grazing from reef organisms and became depleted, much as occurs in the bare "halo" area around wild reefs. Thus, while we had achieved a reasonable simulation of a wild reef, we were having trouble simulating the effect of a neighboring ecosystem, the grass-rich (Thalassia) lagoon, simply because in our system the two components were too close together, so that herbivory by reef organisms on lagoon vegetation was abnormally high.

To correct this problem in our next, much larger (7 kl) system, we included a small (240 l), separate lagoon attached by piping, which prevented access to the lagoon by the larger reef organisms. This lagoon functioned quite well, and for the first time successfully supported the typical lagoon community of (Thalassia) and siphonaceous green algae. While working with this system, however, we noted that while daytime photosynthesis was high enough to achieve high dissolved oxygen levels (7–10 mg/l), dissolved oxygen fell to low levels at night. In addition, concentrations of dissolved nitrogen tended to be abnormally high.

This in turn raised several interesting questions. Shortly before constructing the 7 kl microcosm, several colleagues and I had undertaken an intensive, year-round investigation of the metabolism of a section of reef on the south shore of the island of Saint Croix. We were especially interested in the relationship between geological history, structure (or available surface) and the dynamics of reef metabolism, and indeed were able to demonstrate that geological history had some important effects on reef metabolism (Adey & Steneck, 1985). However, another similarly striking and generally unrecognized factor also came to our attention in the course of this work.

On a diel basis, respiration on typical reefs just about balances photosynthesis, so that there is little apparent loss or gain of energy. This was a feature of reefs that had been generally accepted for about ten years. Nevertheless, the magnitude of respiration is such that, even with moderate atmospheric exchange, the oxygen concentration in the water overlying a well-developed reef can reach very low levels on calm nights – something we observed at Saint Croix on several occasions. Such low oxygen levels probably also indicate comparable degeneration of other water-quality variables such as pH and ammonia concentration, since they are largely the result of high animal biomass. In a typical wild reef, however, constant equatorial and local, wave-driven currents usually maintain ambient water quality by providing a constant supply of high-quality ocean water, day and night. It is only under protected conditions, on the lee side of an island, for example, that water quality is likely to become a factor limiting reef development.

Unfortunately, however, those were exactly the conditions we had in our reef microcosm, and this represented something of a quandary relative to development of improved reef microcosms. At first, we thought of trying to solve this problem by adding a planktonic oceanic simulator to the system. However, the magnitude and cost of what would be required would be high: since the primary production rates of tropical plankton systems are typically low, the system would have to be very large. Of course, photosynthetic rates could be increased by increasing nutrient levels, but this would almost certainly result in lowered water quality. Moreover, continuous scrubbing (i.e., of carbon dioxide and nutrients) with a large volume of planktonic algae would be costly. Next we considered using the reef community itself as a source of high-quality water, in effect operating two interconnected reef microcosms in tandem on alternate light cycles. This, however, would not only have been expensive, it would have been photosynthetically inefficient, since excess primary production in a reef system is relatively small. Finally, I decided to try to remove a piece of the primary photosynthetic component, the algal turf from the reef itself, and to allow that plant community to develop and photosynthesize under appropriately high light intensity in a side branch of the entire system, during hours when the reef was in darkness.

From our experience with both the wild and the microcosm algal turfs, we concluded that we might make this work if, along with the high light intensity, we supplied wave action, water flow, a porous surface (to prevent overgrazing) and constant harvesting (to prevent community succession). Thus, we created a device called "the algal turf scrubber" and attached it to the 7 kl system late in 1979. The algal turf scrubber was extraordinarily successful, in that it achieved primary production rates characteristic of a wild reef, and also simulated the effects of high-quality ocean water by adding oxygen to the system and scrubbing nutrients from it. Most important, it could be operated at night, when water quality is likely to decline, and it was controllable in many ways, since by adjusting light, wave action, water flow and harvest rates we could maintain water chemistry in the microcosm reef much as ocean flow maintains it in the wild. Use of the algal scrubber also demonstrated directly (through harvest of the algae): the unexpectedly high productive capability of algal turfs; the lack of light saturation; the requirement for intensive wave action; and the general lack of restrictive nutrient control over plant production (Adey, 1987). Field testing also corroborated that very high primary production rates are to be expected in oligotrophic seas (with harvestable material averaging 15 g (dry weight)/m²/d, and reaching at least 30 g (dry weight)/m²/d, provided wave and current energy are available to drive the process. Moreover, as a by-product of this understanding we have now developed an entirely new mariculture potential for tropical trade wind seas (Adey, 1987). We have since learned that the same process can be used as a general pollution control device to clean waters of any nature, as long as those waters are capable of supporting benthic algal populations.

Thus, in brief, interactive study between laboratory and field, and a hands-on effort to reproduce a coral reef ecosystem in microcosm led directly to a number of unexpected results. These include: a redefinition of the basis for and the factors controlling primary production in reefs; a sounder understanding of the relationship between benthic and planktonic primary production in the oceans, and the mechanism of nutrient control in these systems; possibly, a way of using the food resources of our oceans much more efficiently; and finally, a technique for removing nutrients from natural waters and sewerage effluents. The microcosm also led us directly to a basic and widely applicable principle of ecosystem function – namely that physical energy input (in this case in the form of waves and currents) can drive primary production largely independent of nutrient supplies or light inhibition. This would have been considerably more difficult to establish in the field, and indeed, although a few authors over the last three decades have mentioned this possibility, no one had seriously considered that the driving of photosynthesis by physical energy could be as important as it now seems. In this case, physical energy input probably operates in several ways: by delivering a constant and non-depletable source of nutrients (albeit of very low concentration) to the plants in the community; by increasing mixing between the aqueous medium and plant surfaces (a wave surge or oscillatory motion would act to break up surface depletion zones more than a uni-directional current); and the constant oscillation of light, a flashing effect on each photosynthesizing cell that allows more efficient utilization of the very high light levels available.

In another series of studies with this reef/lagoon microcosm, we have been able to show that nutrient levels can be driven to high levels quickly (within several days) by adding a heavy animal load in a side system. When we then remove the metabolic load after several weeks, the system recovers to normal levels only asymptotically over many months. This probably results from outwelling of stored nutrients from organic material within lagoon sediments. This recalls the nutrient reservoirs in the sediments of freshwater lakes, which were also identified as a result of manipulative or "synthetic" experiments (see Ch. 8).

Thus, it would appear that transport of material between reef and lagoon must be adjusted to simulate the effects of the one-way relationship that exists in the wild system. The lagoon is an organic sink for the reef. It is also, for some reef organisms, an alternative habitat. While reef workers have generally suggested that a back-flow of nutrients from lagoon to reef through fish (such as grunts) is necessary to maintain reef productivity, this is probably not usually the case since, as described above, algal turfs growing in the wild without significant recycling and under extremely low-nutrient conditions can still achieve very high production levels. Moreover, a generalized lagoon-to-reef water flow (such as usually occurs in reef passes) would probably be detrimental to the coral reef community. Indeed, we have recently arranged a reduced connectivity in our system, thereby achieving sharp drops in nutrient levels without major declines in productivity.

In a microcosm ecosystem, it is difficult partially to isolate lagoon water from the reef system in order to reduce lagoon outwelling to the reef and yet allow the lagoon to function efficiently as a sink for fine reef sediments. We have solved this problem by installing a variety of settling devices throughout the system, chambers designed to remove much of the fine, carbonate sediment that is produced on the reef without removing significant quantities of plankton in the process. In addition, water being returned from the lagoon to the remainder of the system is passed directly through an algal scrubber before returning to the central microcosm unit, the pumping rate being adjusted so that the nutrient content of this water is comparable to that of water ordinarily coming in from the open ocean. In this way, the functional relationship between lagoon, open ocean and reef is maintained by pumps, settling chambers and an algal scrubber.

Finally, in 1980 the 12 kl system briefly described above was modified for exhibit, as well as research (Fig. 9.1). This microcosm included a 1.5 kl lagoon and a reef fish biomass about 50 per cent higher than that of a typical wild reef. In this system the number of algal scrubbers had to be greatly increased to maintain appropriate oxygen, pH and nitrogen levels. In addition, fish predation on some corals greatly increased, so that refugia had to be created for these colonies. Those that are most susceptible are now rotated from the main system to refugia and back periodically, effectively reducing the fish loading effects.

It is important to point out a general principle that my students and I have developed over the years with regard to system simulation: do the job exactly as in the wild if at all practical; otherwise achieve it through engineering. For example, reef waves are wind driven. Wind-driven waves of sufficient magnitude would be impossible to achieve in a small microcosm. Therefore, in our systems, centrifugal pumps provide energy by creating dump troughs directly. More accurate replicas of ocean waves could be created with a pusher-type wave-maker, but this would be more expensive, the pumps would still be needed to provide proper flow, and in any case this level of authenticity has so far proved unnecessary.

Thus, by working with a series of successively modified reef microcosms over a decade, we have considerably improved our understanding of the functioning not only of reefs, but of marine ecosystems generally. Perhaps we could have accomplished this by working with the wild systems alone. However, it certainly would have taken much longer to do so, especially considering our previous understanding – or misunderstanding – of reef systems.

### Other marine microcosm systems

Although I have chosen to describe the development of the Smithsonian's coral reef/lagoon microcosm in some detail, because I have been directly involved in the actual construction of this system and am therefore in a position to give a first-hand account of the value of this process, a number of other marine systems have also been successfully modeled in microcosm. Excellent examples are the temperate planktonic/benthonic system developed at the Marine Ecosystem Research Laboratory (MERL) in Kingston, Rhode Island, to determine the factors influencing the timing of algal blooms in Narragansett Bay (Nixon *et al.*, 1979; Pilson & Nixon, 1980); and a boreal planktonic system created in enclosures in Saanich Inlet in British Columbia, which has been used to carry out trophic studies of this system that would have been difficult or impossible to carry out in the field (Steele & Gamble, 1982; Williams & Le, 1982).

More recently, in our own laboratory, we have built and operated a 7 kl rocky Maine shore coupled with a mud flat and salt marsh, as well as a 40 kl Chesapeake Bay complex with both marsh and open water interconnected systems ranging from tidal fresh to coastal salinity. In cooperation with the Great Barrier Reef Marine Park Authority, we have also completed a 3000 kl reef mesocosm in Townsville, Australia, that promises to teach us much about larger systems (Kelleher, 1986).

### Practical considerations

Marine microcosms of significant size are relatively costly, perhaps generally costing about as much as major pieces of electronic laboratory equipment. However, marine field work, especially on a research vessel or on a distant shore or island, is also costly and can be a major budgetary element in marine research. While it might appear that terrestrial microcosms would be simpler and less expensive than marine systems, this may not be the case if soil, ground water table, wind and lighting is considered, especially if the system being modeled does not belong to the local climate. Microcosms in general are probably best regarded, like other expensive instrumentation used in basic research, as well worth the expense. They provide alternate views of the often unmanageable real world, and also alternatives to more theoretical, but heavily human-biased, approaches. From the more purely practical point of view of marine ecosystem use in restoration and management, it seems likely that the value of microcosms would be comparable to that of models in harbor and hull design – a way of ensuring that the worst mistakes are made on a small rather than a large scale.

### Conclusions

Carried out in conjunction with extensive field investigations and, where appropriate, with mathematical modeling, microcosm studies can provide insights into complex communities that might not otherwise be available. While microcosms are not simple, and scaling and systems interaction may make them in some ways even more intractable than their wild counterparts, by their very nature they are easily manipulated. Microcosms provide full experimental access to ecosystems and bring ecology into the same frame of reference as the laboratory sciences. On the other hand, while conceptual/mathematical models are essential components of ecosystem research, they can only be as good as the modeler's concepts, and are capable of providing contentment in total error.

Scaling is without question the basic problem of microcosm modeling. On the other hand, the scaling effects of microcosms and mesocosms can be accounted for, and can even be used to great advantage in studying wild systems and exploring questions about scaling that are of great theoretical interest (see Ch. 19).

In more practical terms, the rebuilding of wild ecosystems might well be approached as a way of researching basic questions with a series of successively larger microcosms or mesocosms. These systems would effectively be prototypes or pilot systems, and the actual rebuilding of the ecosystem would then be an upscaling of this work by considerably better informed ecologists. This approach places the ecological restorationist in the right philosophical frame of reference, in which the system is regarded not in terms of an idealized, over-simplified model, but as a complex piece of machinery that one handles, interacts with and only partly understands. Just as in education the value of didactic and heuristic modes of instruction are endlessly debated, so one can debate mathematical as opposed to physical modeling with almost the same types of arguments. However, as the results of self-education can be strikingly successful, so microcosms have an important role to play in our efforts to understand and repair complex ecosystems.

### References

Adey, W. (1983). The microcosm: a new tool for reef research. *Coral Reefs*, 1, 193–201.

Adey, W. (1987). Food production in low nutrient seas. *Bioscience*, 37, (5).

Adey, W. & Steneck, R. (1985). Synergistic effects of light, wave action and geology on coral reef primary production. *NOAA Symposium Series for Undersea Research*, 1 (2), 163–87.

Basta, D. & Moreau, B. (1982). Introduction to analyzing natural systems. In *Analyzing Natural Systems*, ed. D. Basta & B. Vower, pp. 23–96. Baltimore: The Johns Hopkins University Press.

Bradbury, R. (1977). Independent lies and holistic truths: towards a theory of coral reef communities as complex systems. *Proceedings of the 3rd International Coral Reef Symposium*, 1, 1–8.

Gardner, H. (1985). *The Mind's New Science: The Cognitive Revolution in the Computer Age*. New York: Basic Books.

Goldberg, E. D., McCave, I. N., O'Brien, J. J. & Steele, J. H. (eds.) (1977). *The Sea, Marine Modeling*, Vol. 6. Bristol: John Wiley & Sons.

Grigg, R., Polovina, J. & Atkinson, M. (1984). Model of a coral reef ecosystem. *Coral Reefs*, 3, 1–27.

Henderson, R., Smith, S. & Evans, E. (1976). *Flow Through Microcosms for Simulation of Marine Ecosystems*. Hawaii: Naval Undersea Center.

Hudson, R. (1979). *Coastal Hydraulic Models*. US Army Corps of Engineers Special Report 5.

Kearns, E. & Folsome, C. (1981). Measurement of biological activity in materially closed microbial ecosystems. *Biosystems,* **14,** 205–9.

Kelleher, G. (1986). Managing the Great Barrier Reef. *Oceans,* **29** (2). *The Great Barrier Reef: Science and Management.* American–Australian Bicentennial Issue.

Lewis, J. (1977). Processes of organic production on coral reefs. *Biological Reviews,* **52,** 305–47.

Nixon, S., Oviatt, C., Kremerm, J. & Perez, K. (1979). The use of numerical models and laboratory microcosms in estuarine ecosystem analysis – simulation of a winter phytoplankton bloom. In *Marsh–Estuarine Systems Simulation,* ed. R. Dame. Carolina: University of South Carolina Press.

Perez, K., Morrison, G., Lackie, N., Oviatt, C., Nixon, S., Buckley, S. & Heltshe, J. (1977). The importance of physical biotic scaling to the experimental simulation of a coastal marine ecosystem. *Helgolander Wissenschaft Meeresunters,* **30,** 144–62.

Pilson, M. & Nixon, S. (1980). Marine microcosms in ecological research. In *Microcosms in Ecological Research,* ed. J. Giesy, pp. 724–41. Technological Information Center of the US Department of Energy.

Steele, J. & Gamble, J. (1982). Predator control in enclosures. In *Marine Mesocosms,* ed. G. Grice & M. Reeve, pp. 227–37. Berlin: Springer–Verlag.

Stuntz, G. (1963). The use of model basins in the design process. In *Chesapeake Sect.,* Society of Naval Architects and Marine Engineers, pp. 1–14.

Williams, P. J. & Le, B. (1982). Microbial contribution to overall plankton community respiration-studies in enclosures. In *Marine Microcosms,* ed. G. Grice & M. Reeve, pp. 305–21. Berlin: Springer–Verlag.

Williams, S. & Adey, W. (1983). *Thalassia testudinum* Banks ex Kinig: seedling success in a coral reef microcosm. *Aquatic Botany,* **16,** 181–8.

*Michael E. Gilpin*
Department of Biology, University of California, San Diego

# 10 Experimental community assembly: competition, community structure and the order of species introductions

The challenge of actually restoring ecological communities raises certain questions that go beyond, and in some ways are more complex, than those encountered in attempts merely to reestablish some form of biological function. In community restoration the focus is on species populations rather than on such "lumped" quantities as the biomass in trophic compartments or the flows of materials and energy between compartments. Thus, questions arise concerning not only the species that should be used, but also the size of the introduced populations and the sequence in which they should be added to the community.

In a sense, the best way to answer these questions is actually to carry out the restoration. If experiments are to produce information of general, rather than merely local or empirical value, however, they must be carried out under controlled conditions and with appropriate controls and replications; this may be difficult or even impossible to accomplish in the field (see, for example, Ch. 13). Fortunately, however, it is possible to carry out experiments involving assembly of small-scale communities designed specifically to test ideas and to establish or refine *principles* relevant to community restoration.

In the preceding chapter, Walter Adey described work involving construction of mesocosms of various marine ecosystems. That work, however, was largely ecosystem- rather than community-oriented, and focused for the most part on various aspects of system function. In this chapter I will describe a series of experiments that are community- rather than ecosystem-oriented and that involved experimental assembly of highly simplified communities under rigidly controlled conditions in order to identify crucial factors influencing the structure and dynamics of communities generally, and also to test the consequences of different orders of species introduction, whether these result from chance in a natural succession or from human choice in managed restoration. Of special interest here are the opportunities this synthetic approach with a highly simplified community offers for isolating factors, such as competition, that underlie changes in communities, but that may be obscured in more complex systems, or in systems exposed to changes in weather or other stochastic events. While the direct applicability of experiments of this kind to the problem of actual restoration in the field may be questioned, it would seem that isolation and characterization of these factors is crucial, not only to basic ecological understanding of the communities in question, but to the development of principles (as opposed to mere empirical prescriptions) for their restoration.

The experiments I will describe in this chapter were indeed carried out with communities stripped down to the minimum essentials and representing natural communities only in a highly abstract sense. These were mixtures of species of fruit flies (*Drosophila*) living over extended periods in quarter-pint milk bottles, and in a sense may be regarded as a series of carefully controlled experiments in island biogeography. As in the descriptive research and thought experiments often associated with island biogeography, we began with a list of potential member species, $P$, from which we selected species for introduction into the bottles in various sequences and combinations, rather as species might be expected to reach a depauperate island over a period of time. This species pool can be conceptualized mathematically as a state space of $P$ dimensions, with an axis for each of the species introduced (Fig. 10.1). The population of each species introduced may then be plotted over time, the resulting series of points defining the state or configuration of the community as it develops in time. A line connecting these points in the order in which they occur records the history of the successional assembly and is customarily called a "trajectory".

In Chapter 2, Bradshaw used the idea of a trajectory to describe succession and restoration in general terms. Although Bradshaw's state space is purely metaphorical (since he assigns his axes no explicit ecological meaning) he nevertheless makes several points pertinent to this discussion. The initial segment of Bradshaw's trajectory represents changes in a natural ecological community resulting from disturbance. This trajectory is represented by a single line, or by lines that tend to converge, since the effect of disturbance is to simplify the community by making prevailing conditions increasingly harsh and reducing the number of species present towards zero. From this point on, however, as the system enters what might be called the recovery or restoration phase, a number of different trajectories are possible. Of these, it is the fourth that is of most interest to restorationists, since it is this trajectory that represents the actual recovery – or restoration – of a community that matches the original or model community, not only in function, but in species composition. Like the downward phase of the curve representing degradation, the restoration trajectory may have a single, ideal endpoint, and can also reach this endpoint in several different ways. It can, for example, go slowly or rapidly. It may go directly or circuitously. Finally,

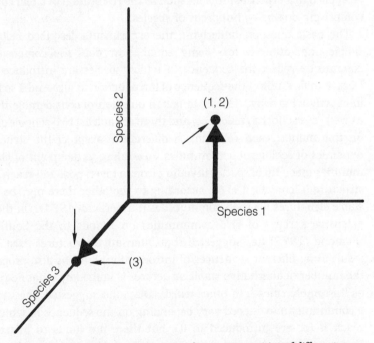

*Fig. 10.1.* Graphic model suggests how communities of different composition may result from a bringing together of the same species in different sequences. In the first case (1, 2) species 1 and 2 are introduced in sequence and the two coexist and reach equilibrium in a community that is resistant to invasion by species 3. In the second case (3), species 3 is introduced first and proves resistant to invasion by either 1 or 2.

it may proceed through a definite sequence of species, or it may follow a modified sequence, or may even skip species altogether. In either case, it is with this fourth pathway that the restorationist is most concerned, and where the restoration ecologist can best demonstrate his or her knowledge and worth, since it is the ability to control this process with confidence that is not only the objective of restoration, but also, as Bradshaw argues, the "acid test" of ecological understanding in the most basic sense.

Bradshaw's diagram is fine for illustrating these issues in a general way, but a more exact representation is necessary if we are to deal effectively with the details of the ecological (and eventually the economic) aspects of the problem. As it happens, the P-dimensional representation of species densities within a community described above is sufficiently exact for this purpose, and it will be used in the following discussion. A caveat is, however, that this representation may, in fact, be *too* detailed, and that some sort of amalgamation of axes would be necessary if this mode of reasoning were to be applied to a practical problem such as the restoration of a real community comprising dozens or hundreds of species.

The basic question underlying the experiments described below was a simple one: other factors being equal, how does the composition of a community reflect the sequence in which species are introduced into it? Does it, in fact, reflect the sequence of introduction at all? And if so how, and in exactly what way? This is, in fact, a question of considerable theoretical as well as practical significance, and there are at least two schools of thought on this matter, each reflecting a different concept of the structure and dynamics of ecological communities. According to one point of view, communities made up of species invading from a given pool will reach a unique structure (Clements, 1916); according to the other there may be alternate stable structures for such communities (Sutherland, 1974). On the basic of extensive surveys of bird communities on islands in the South Pacific, Diamond (1975) has suggested that alternative structures exist and may result from different sequences of introduction, but has also proposed that the number of alternative stable structures is limited by what he referred to as "assembly rules". In other words, Diamond suggested, the structure of a community may indeed vary depending on the sequence in which species reach it (or are introduced to it), but there are limits to this variation. Variation occurs, but within certain patterns. Attempts to resolve this issue and to define the proposed assembly rules by descriptive studies have been inconclusive, however, because of the field biologist's inability to control background factors. While in principle it ought to be possible to resolve these factors by statistical analysis of the data collected by observation, doing this has proved very difficult. For discussions of this issue, see Connor & Simberloff (1979) and Gilpin & Diamond (1982). In fact, the real test of Diamond's rules is the actual synthesis of the competition system. The experiments described here were undertaken, therefore, partly as a test of the validity of Diamond's assembly rules.

The reasoning behind these experiments is as follows. Suppose that we have 26 hypothetical "species" to introduce into a system, and that we denote these species A, B, C,..., Z. These species or elements may be introduced to the community in various sequences for example:

(A, C, D, Z, ..., H) *or*
(C and D and G, Z, W, ..., K) *or*
(Z, W, A and B, C and D, ..., F)

where the leftmost element is introduced first, and where the "and" signifies simultaneous introductions. After the introductions and subsequent adjustments and sorting out, which are mediated by such ecological forces as competition, predation, and so forth, the system will come to a final state, which we denote by listing the species or elements present in alphabetical order inside angle brackets. For example:

⟨A, B, F, M, W, Z⟩ *or*
⟨G, H, T⟩.

Note that we are talking here in very abstract terms, and that this final configuration of elements (species) may be a physical structure such as a house or a violin, a biological structure such as an ecological community, or even a psychological structure such as knowledge or a career. In any case, there are a number of relationships that may exist between the final state and the sequence in which a given set of species or elements is introduced to the system. First, given a certain set of conditions and a certain pool of species or elements from which to draw, it may be that only a single final state is possible, and that this final state is completely independent of the sequence in which the elements are introduced to the system. Alternatively, the final state may be strictly dependent on the order of introductions, each sequence producing a distinctive, characteristic result. Or the relationship between the introduction sequence and the final result may be somewhere in between these two alternatives – and more complex than either or them – with various introduction sequences leading to a limited number of *alternative* stable structures.

A few examples from real life may help to clarify these distinctions. A person's knowledge is almost completely independent of the order in which he or she studies calculus and Spanish, but strongly dependent on the order in which a person studies calculus and mechanics, since without calculus the progress one may make in mechanics is severely limited. Similarly, the color of paint is independent of the order in which pigments are added, while in building a house it is necessary to lay the foundations before erecting framing, and so forth.

The situation with ecological systems, however, is not so clear cut. That is, in some respects it is strictly necessary in assembling such a system to follow a certain sequence of introductions. In fashioning ecosystems such as Adey's marine mesocosms, for example, it is obviously necessary to add the substrate, the water and the plants before adding the fish. The reason for this is that at this level the system is like a house or a violin, in that it has a definite structure that must be achieved for the system to function. Indeed, the extent to which the final product depends on a given sequence of additions is directly related to the extent the system itself has what might be called structural integrity. Thus the assembly of the system, and in particular the exploration of various sequences in the assembly procedure, offers a way of probing and characterizing the nature of this integrity.

The question now is, to what extent and in just what ways are ecological systems integrated? Clearly, at the level of broad ecosystem function, such as that implicit in the example of the marine mesocosm noted above, there is a good deal of integration, a working together of parts that is necessary if the system is to function – and to exist at all. The question becomes more complex, however, at finer levels of organization within the system regarded as a community. Are there, for example, alternative stable communities of, say, predators, guilds of competitors or communities of mutualists?

Here the answer is not so obvious, yet it is this question that is most directly related to the problems encountered in attempts to reassemble ecological communities. In these terms the question is whether, starting out with the same list of species, and holding conditions constant, one introduction sequence will result in a community with species list $\langle A, C, E, \ldots \rangle$, while another will result in a community with quite a different set of species – say, $\langle B, D, F, \ldots \rangle$ – or whether the same species will add up to the same community regardless of the order in which they are introduced.

Clearly, in the absence of major changes in conditions, we are not likely to wind up with two completely different species lists. Yet the composition of various subsystems within the community may differ, and this may be of great importance in the case, for example, of a restoration effort carried out partly to provide habitat for a particular rare species. In any case, it is variations in structure within these various subsystems that are at issue here, and in this chapter I will consider only one of these variations – the guild of interspecific competitors, the same unit of organization discussed in Chapter 13.

A key point here is that the experimental, synthetic approach offers certain advantages in studies of this question. Attempts have also been made to resolve the question of alternative community structure on the basis of descriptive studies carried out in the field. Connell & Souza (1983) recently reviewed attempts to do this, however, and concluded that the results are inconclusive because it is always impossible in studies of this kind to rule out or to account for variations or changes in the biotic or abiotic background. That is, even if one finds what appear to be alternative stable communities that have developed in different areas, these may reflect subtle variations in the environment. Actually, these authors did not discuss the set of field data that perhaps offers the strongest argument for alternative community structures resulting from different assembly sequences. This is the "checkerboard" distribution of species pairs on islands, described by Diamond (1975). In this pattern, one of two species will be present on all the islands of an archipelago, but the two species are never found together on the same island, even though in the absence of any organizing force they would be expected to co-occur frequently. For example, if each species occurs on half the islands in an archipelago, then, on purely statistical grounds, one would expect to find the two together on about a quarter of the islands. Since this is not the case, Diamond speculated that the association of these species is not random, but is subject to some organizational influence. He understood, however, that in order to test this idea rigorously with these communities it would be necessary to carry out perturbation experiments that would probably be illegal or unethical, if not impossible. In fact, it seems that this issue cannot in principle be resolved by field studies alone, since, although communities may indeed differ from location to location, it is never possible to be certain that these differences are not due to variations in conditions at the two sites.

There is, of course, a second approach. As Adey pointed out in his discussion, mathematical modeling can play an important role in attempts to identify the critical factors governing the behavior of a system. Indeed it is a routine exercise to write down equations such as the Lotka–Volterra equations (Lotka, 1925) that describe at least some of the essential aspects of the dynamics of a community, and it is easy to show that these do have alternative equilibria, that they depend on the sequence of species introductions, and that some combinations of species depend on the earlier presence of other species that may not be present in the final, equilibrium community (Gilpin & Case, 1976). A simple example, suggesting how such alternative equilibria might develop, is illustrated in Figure 10.2. Here it is assumed that there are four species, each of which utilizes a distinctive zone of a two-dimensional resource continuum, some of which overlap. Here the sets of species ⟨A, C⟩ and ⟨B, D⟩ coexist readily because their patterns of resource base use do not overlap. In contrast, combinations such as ⟨A, B⟩ do overlap and thus form unstable combinations, at least one member of which is likely to decline to extinction (Gilpin, 1975).

Such conceptual demonstrations have great plausibility, and indeed there can be little doubt that an internal influence, such as (in the case above) interspecific competition, does occur and, at least in theory, might play some role in shaping a developing community. The question is, however, just how powerful are these influences? That is, to what extent do they actually contribute to the shaping of the community, and to what extent (or under what conditions) are they minor factors that are obscured by other factors, such as, for example, stochastic factors? To answer this question it would seem that neither descriptive field studies nor conceptual models are enough. What is needed is an experimental investigation involving actual assembly of communities.

There are some stringent requirements for these experiments, however: for one thing, the pool of potential invaders must be fairly large; for another,

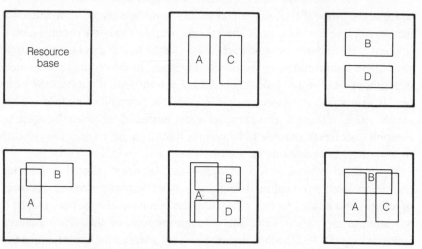

*Fig. 10.2.* Several species depending on the same resource base may coexist if they exploit the resource in different ways and so do not compete directly. Thus for the four species, A, B, C and D, shown here, a number of combinations are ruled out, but the two pairs, ⟨A, C⟩ and ⟨B, D⟩, that can coexist represent alternate stable combinations.

since a large number of combinations must be run, with suitable controls and replicates, the system must be small and inexpensive, and it must develop rapidly; finally, the physical background and biological resources must be rigidly controlled and standardized.

As it happens, a set of *Drosophila* species that can "invade" or be added to the standard quarter-pint media bottles routinely used by geneticists for studies with *Drosophila* fitted the bill perfectly, and this was the system we chose to carry out the experiments suggested by the preceding discussion. Briefly, we first selected five *Drosophila* species of roughly equal competitive ability, as assessed in a series of two-way "tournaments" in the media bottles. We then allowed these to colonize bottles in different sequences. Altogether, we used ten different sequences, adding 13 male/female pairs in each case, with a four–week interval between introductions of successive species. Four weeks is roughly two *Drosophila* generations and, given the high intrinsic rate of increase of these flies, allows for saturation of the environment. Thus, all introductions were against entrenched competitors.

Each sequence was repeated twice and the results of the two trials were compared. The concordance between them was about 90 per cent, indicating that it was species biology, not chance, that was driving the system.

The results of these experiments are described in detail elsewhere (Gilpin, Carpenter & Pomerantz, 1986). Here I will discuss the results for only a single subset of five species. One of these I will ignore, since it was never present in the final, equilibrium community. To avoid unnecessary drosophilisms, I will refer to the remaining four species as PAUL, ZAP, NEB and ANB ("an-bee"). What we found was that these four species can indeed form three alternative equilibrium configurations or "communities", depending on the sequence in which they are introduced into the system. First, the system can go to a "monoclimax" of all PAUL or all ZAP; or it can reach a state in which NEB and ANB coexist.

It is possible, moreover, by a careful analysis of these systems, to infer the mechanisms underlying them. Basically, it is possible to account for all three "permitted" combinations on the basis of preemptive competition (Yodzis, 1978) between the species. If, for example, PAUL establishes itself at saturation density (about 500 adults and 5000 larvae per bottle) no other invading species population of 26 individuals is able to establish an ecological beachhead. It is clear, moreover, that the reason for this is the reproductive behavior of the female fruit flies. The females simply have too much trouble finding mates of their species when they are present at such a low density. In other words, the rate of insemination and egg laying depends on the "concentration" of the species in the mixture. this implies a positive feedback system, with the low frequency population reproducing more slowly than the dominant species, and so declining even further. With death rates of adults running at roughly 30 per cent per day in these systems, a delay in insemination of as little as six hours can result in a reduction in a competitive disadvantage of as much as 10 per cent.

What was most significant about these experiments, however, was that the patterns of community structure for our fruit fly communities turned out to be remarkably similar to those Diamond had proposed on the basis of his studies of bird communities in the field. This constitutes a verification of the hypothetical assembly rules, which were impossible to verify on the basis of the data collected by observation alone. It also implies that, at least for close ecological analogues such as competitors for the same simple laboratory habitat, competition is indeed a significant factor in the shaping of a community.

Two points remain to be emphasized. The first is that it would have been impossible to reach this conclusion without actually assembling experimental communities under carefully controlled conditions. Here the synthetic process allowed us to study how this factor actually influences the coming together of the community. Carrying it out with a relatively simple community under rigidly controlled conditions in the laboratory allowed us to isolate this factor from other influences that might have obscured its operation under field conditions.

The second point is that the results clearly have implications for restoration, since they indicate that, at least under certain conditions, the composition of an artificial community may depend on the order in which component species are introduced to it. One might not think it would matter which of two or three species of fruit flies one wound up with in a restored community, but of course ecological communities have a multi-layered trophic structure, so that a seemingly small change in one level might produce a cascade of effects in other levels.

In such cases, it is important that great care be taken to identify the crucial ("keystone") species and to introduce them in the correct sequence. At the same time, an important question remains. Since these experiments were carried out with relatively simple communities, under controlled conditions, specifically in order to make it possible for us to discern the role that competition and other factors intrinsic to the community play in its development, it is not clear from these experiments to what extent and under what conditions these factors actually influence the pattern of development under field conditions. To determine this, it will be necessary, in effect, to extend these experiments into the field and to test similar hypotheses in the course of actual restoration experiments. Indeed, by proceeding in this way, we may hope to discover the hierarchical levels at which various factors such as competition have the greatest influence on the shaping and dynamics of the community. Even if competition, for example, plays a crucial role in stripped down communities such as our *Drosophila* communities, it is not clear just how it influences events in more complex communities under field conditions. The prairie restoration experiments proposed in Chapter 6 represent a step in this direction. In general, it is clear that this is a venture that the theoretical ecologist must entrust to colleagues operating in the world of biological reality.

### References

Clements, F. C. (1916). *Plant Succession*. Washington DC: Carnegie Institute.

Connell, J. H. & Souza, W. P. (1983). On the evidence needed to judge ecological stability or persistence. *American Naturalist*, **121**, 789–824.

Connor, E. F. & Simberloff, D. (1979). The assembly of species communities change or competition? *Ecology*, **60**, 1132–40.

Diamond, J. M. (1975). The assembly of species communities. In *Ecology and Evolution of Communities*, ed. M. Cody & J. M. Diamond. Cambridge: Bellnap & Harvard University Press.

Gilpin, M. E. (1975). Limit cycles in competition communities. *American Naturalist*, **109**, 51–60.

Gilpin, M. E., Carpenter, M. P. & Pomerantz, M. J. (1986). The assembly of a laboratory community: multispecies competition in Drosophila. In *Community Ecology*, ed. J. M. Diamond & T. J. Case New York: Harper & Row.

Gilpin, M. E. & Case, T. J. (1976). Multiple domains of attraction in competition communities. *Nature*, **261**, 40–2.

Gilpin, M. E. & Diamond, J. M. (1982). Factors contributing to non-randomness in species co-occurrences on islands. *Oecologia* (Berlin) **52**, 75–84.

Lotka, A. J. (1925). *Elements of Physical Biology*. Baltimore: Williams & Wilkins.

Sutherland, J. P. (1974). Multiple stable points in natural communities. *American Naturalist*, **108**, 859–73.

Yodzis, P. (1978). *Competition for Space and the Structure of Ecological Communities*. Berlin: Springer–Verlag.

# IV | Partial or piecewise restoration in the field

> The test of whether we truly understand such a (complex) system is no longer our ability to predict it, but our ability to construct another system that does the same sort of thing as the original.
>
> Frederick Turner, *The Predicting Contest*

One reason why ecological restoration may not seem to lend itself to systematic experimentation is that it typically involves the wholesale assembly of systems by the bringing together of numerous elements more or less simultaneously. Since experimentation involves the deliberate modification of single factors while holding all others constant, or, as far as possible, accounting for them through appropriate controls, it may be that the value of such wholesale assembly of systems is extremely limited.

What this would seem to suggest, however, is the piecewise assembly of systems, not *en masse*, but one element at a time. Indeed, this approach is frequently employed by ecologists, and might be construed as a linking of the laboratory synthesis carried out for explicitly heuristic purposes with the more "realistic" work of restoration in the field. This, in other words, might be what restoration ecology would look like if it were developed systematically as an approach to ecological research.

What this amounts to in one sense is the extension of techniques now being used for studies at the population level to work with whole communities. The chapters that follow provide some sense of this approach and its value in dealing with a variety of ecological issues. (Additional research pertinent to this idea is also described by Pat Werner in Chapter 22.)

*Eugene S. Morton*

Department of Zoological Research, National Zoological Park, Smithsonian Institution, Washington DC

# 11 | Reintroduction as a method of studying bird behavior and ecology

If a tropical forest were cut, burned, planted to grass, and grazed by cattle we would not expect to learn much by releasing forest birds back into the resulting treeless terrain. We might expect to learn a great deal, however, if the habitat were more suitable, even if it had been reduced in size or otherwise disturbed in some way. In fact, by reintroducing a single missing bird species, we would be performing a "one-piece" restoration of the pre-existing community. And, even if the species did not survive, we could learn much about both the species itself and its specific habitat requirements (see Long, 1981).

Reintroduction is a way to study animals that presents us with unique opportunities to learn more about them. For example, the released animals have the opportunity to "choose" habitats without constraints resulting from the presence of conspecifics. For managers, the criterion of a successful reintroduction is successful reproduction by the reintroduced species. But reintroduction can tell the biologist a great deal about a species even if reproduction is not achieved. For example, if the introduced individuals or propagules survive for only a few days, their decline may reveal much about the species' requirements and the causes of its original extirpation.

Reintroducing species is a much more complex problem than the mere mechanics of the process might suggest. Most of today's avian ecologists are simply not prepared to consider the cues used by individual birds in their "personal" attempts at survival and reproduction. Unlike those ecologists who study large mammals or primates, and especially those who study adaptations of individual animals within a social group, ornithologists tend to assume that "a bird is a bird" and can be studied on the species level rather than on the individual or population level. The study of avian communities tends to lead us away from studying those adaptations of individuals with which we should be concerned. This is very limiting because species actually exist as populations of individuals, and, in birds as in many other groups, variations within populations are often ecologically significant. In fact, one way to demonstrate this and to study it is by trying to reintroduce the species to an area from which it has been extirpated. Such attempts force us to study the adaptive nature of individuals in relation to their habitat. Actually, many topics of interest can be approached in this way. Here, for example, is a list of some general topics for which data are available or might be obtained through reintroduction. Though expressed simply as subjects of interest, each of these could be restated as specific hypotheses to be tested by appropriately designed reintroduction:

1.  Habitat selection and use
    a.  foraging microhabitat and effects of food availability
    b.  nest placement strategies
    c.  cues used in habitat selection (overall gestalt or more specific cues?)
    d.  importance of presence or absence of interspecific competitors and socially dominant species
    e.  territory or home range size in the absence of intraspecific competition
2.  Social behavior
    a.  mate choice and pairing behavior; ability to find mates when population density is low
    b.  breeding behavior and period of adult/young relationships
    c.  response to predators or disturbances
3.  Other
    a.  demographic profile of a known population
    b.  causes of original extirpation: disease, seasonal food constraints, predation, parasitism, rare necessary microhabitat or serial stage of habitat, habitat area effects
    c.  contrasts in survival, habitat use and social behavior be-

tween wild-caught and captive-bred propagules and factors responsible for these differences.

I would like to discuss some of these points in an example, a reintroduction study I did on Barro Colorado Island (BCI) in the Panama Canal. BCI is roughly circular and about 15 km² (6mi²) in area. It was formed when Gutan Lake and the Panama Canal were dug in 1913. Originally, BCI was a hilltop covered and surrounded by forest. Although it is an island, it is important to keep in mind that it still is, in effect, a hilltop. Since 1923, it has been protected from legal hunting and has been studied extensively. The Smithsonian Tropical Research Institute maintains it to this day, with cooperation from the Panamanian Government. Fortunately, several people, notably Willis & Eisenmann (1979), have completed long-term studies of bird populations on the island. They found that between 1960 and 1971, 42 species of birds were extirpated from BCI (Willis, 1974a, b). Of these, most were lost due to the maturing of the island's forests. (The north-east half of BCI was not forested at the time of isolation and supported many non-forest species.) But at least 14 species seemed to have died out for various "nonsuccessional" reasons. These were forest birds that had plenty of natural habitat on BCI. More recently, Karr (1982) estimated that as many as 50–60 forest bird species may have been lost from BCI. Karr's estimate is based on a comparison between species on a nearby mainland site and those on BCI today. A theory offered to explain this die-off, the "island biogeographic theory", was prevalent at the time of Karr's study. This theory suggested that the small size of the island would of itself result in extinction of a certain number of species – that in fact one could predict within a certain variance how many species might be lost from BCI after it became an island, simply by knowing the size of the island. Karr, however, suggested an additional factor, peculiar to BCI itself, to explain the situation. This was the fact that BCI does not provide a sufficient mosaic of habitats to enable some species to survive certain climate-induced bottlenecks. On the mainland, Karr had found, many species move to more mesic habitats during severe dry season conditions, and BCI may not offer microhabitats sufficient to permit this escape. Certainly the island biogeographic theory did not explain what had actually occurred. In any event, the past work by Willis, Eisenmann and Karr set the stage for my "synthetic" reintroduction experiment.

I wanted to know the immediate causes of extirpation for at least some of the species lost from BCI because I wanted to explore the feasibility of reintroducing species to forest preserves in the tropics. By selecting reintroduction as a means of achieving this, I felt I would simultaneously be carrying out a pilot reintroduction and testing a number of hypotheses concerning the causes of extirpation. If, for example, the conditions that caused the original extirpation had lasted only a short time, we would expect the reintroduced birds to live about as long as individuals of that species lived before they disappeared, and we should also see some reproduction. On the other hand, if the conditions are chronic, then we might not expect reproduction and the lifespans of the introduced birds could be used as an index to how "bad" conditions are and might also provide clues to help us solve the problem.

Because I had already been studying wrens, I decided to reintroduce two species of wrens selected from the list of extirpated species: the song wren (*Cyphorhinus phaeocephalus*) and the white-breasted wood wren (*Henicorhina leucosticta*). Both species were common on the mainland, but apparently would not cross even a small, 300 m watergap to recolonize BCI. Both are typical in several ways of many other species. They are insectivorous, and their social systems are of the most common type. Like about 60 per cent of all the perching birds of Panama, they form permanent, year-long pair bonds, and maintain permanent territories. I thought that their sedentary habit might facilitate relocating them after release. Presumably, they would not quickly fly away, but that remained to be seen.

I captured seven wild individuals of each of the two species on the mainland and immediately carried them back to BCI, where I released all of them from the same location. The number seven was not cast in stone, but it did permit the release of birds captured both as breeding pairs and as singletons. I kept track of pairs by using colored leg bands. I did not expect this reintroduction to be successful – that is, I did not expect that the birds would reproduce and become once again a normal part of the avifauna of BCI. They had been extirpated there once and probably would be again. What I wanted was to find out what had caused their demise, and I expected that I might learn something about this by observing my introduced population as it became extinct.

I released the wrens in June 1976, and checked on them the following December, throughout the following May, and then every October until 1980, by which time all but one of them had apparently died. This last individual may still survive on BCI today, since it was last seen in December, 1985.

When I first checked these populations, six months following reintroduction, I was interested in testing a number of hypotheses related to the near-absence of social or territorial pressure. Would the reintroduced birds settle in the best habitats available because of this lack of pressure? Would pairs break up and then reform easily, or would the birds' territoriality and permanent pair-bonding affect their ability to find one another? (This turned out to create no difficulty, as the birds generally formed into new pairs following reintroduction.)

In December 1976, using taped playbacks of their songs to stimulate a vocal response, I found that two pairs of song wrens had swapped mates and were living 2000 m apart. One pair was right at the top of the island and the other was very near the release point. Both of these pairs, moreover, had successfully fledged clutches. A bird that was extirpated on the island had, therefore, not only managed to survive but had also reproduced – a sure sign that this reintroduction was successful.

The birds seemed to be doing splendidly. At the time, I could not figure out why the birds had settled so far apart, for there was intervening forest habitat that seemed to me to be equally capable of supporting them (Morton, 1978).

One hypothesis that I tested in an attempt to explain this situation derived from competition theory. This was the idea that the song wrens would choose territories where potential competitors – in this case antwrens and antbirds – were least common. Since the original demise of the wrens on BCI, the antbirds had become more common there than on the nearby mainland. One could encounter a flock of antwrens about every 200 m along the trails on BCI, whereas on the mainland, one met them only about every 1000 m. I thought, due to their abundance on BCI, the antbirds might be competitors of the wrens and might even cause their extinction. However, this was not the case. I found that the wrens had settled in locations bearing no relationship to the density of antbirds.

Interestingly, I also found that the white-breasted wood wrens had chosen habitat that was, in fact, not forest but little patches of second growth left on the island's shoreline and in large tree-falls. I concluded that the earlier loss of that species was probably due to a loss of habitat as the island's forests matured. So their demise was not a "nonsuccessional" loss as had been suggested after all.

In contrast, the song wrens had settled in forest habitat, and they weighed at least as much after they were reestablished on BCI than when I had captured them. This ruled out any chronic problem of obtaining food. The possible cause of their earlier extirpation gradually narrowed. I finally became convinced that song wrens had become extinct on BCI because of the type of nests they build and where they place them.

I had found the nests of the first two song wren pairs that had successfully fledged young. These nests had been placed near the most sluggish streams on the island. These nests, and two others of the seven I eventually found, had been built in saplings of one tree species (*Psychotria grandis*), and all were about 1.7 m from the ground. The nests consisted of a loose heap of leaves and long leaf petioles and mimicked a bunch of flotsam caught in the crotch of a sapling during high water. As a stream rises and falls periodically, it leaves many such clumps of deritus trapped in the vegetation along its course. The song wren nests bear a striking resemblance to these clumps, and were even oriented downstream, with longer, hanging stuff pointing in the downstream direction.

The significance of this seems obvious. The song wren not only uses its well-camouflaged nest for breeding but, like many tropical species, spends every night there; their nests are dormitories as well as nurseries. If your nest itself is rare relative to the detritus piles that it mimics, this may be an effective ruse for avoiding predators, especially for a species that is relatively rare. Unfortunately however, because BCI is a hilltop it has very few streams that are sluggish and slow-moving. In fact, it has none that rise and fall to the extent that lowland streams do. The first two song wren nests were successful, I surmise, because they were placed along streams. These were temporarily successful, but eventually they were discovered by predators and the birds were frightened out of them and looked elsewhere for new nest sites. Surprisingly, all of the nests we found following this initial period were along trails. Why? The trails apparently looked to the birds like streams, and, in fact, some BCI trails do carry water during heavy rains. Consequently, I suggest that the apparently slight habitat alteration on BCI caused by the creation of miles of trails could have led to the demise of the song wrens.

The problem is probably that trailsides make poor substitutes for the banks of a sluggish stream as nesting habitat for song wrens. Appropriate streams are rare on BCI in the first place, and song wren populations were always probably low there. Creation of trails, however, resulted in the appearance of an alternative habitat which, though apparently attractive to the wrens, is not really suitable for nesting, at least partly because the wrens' nest camouflage strategy is unsuccessful along trails. Many predators do use these trails, and I found evidence that predation was a problem for birds nesting in these sites. On 20 October 1978, the nest of the last breeding pair, which consisted of an introduced female and her surviving son, was found pulled apart. The female had been killed. Today, only a single song wren remains on BCI and lone birds do not build dormitories.

This brings us back to the question of a species' critical adaptations and the specific attributes of its habitat. In the case of the song wren, a critical element in the wren's life cycle was the permanent dormitory, a nest that mimics detritus and that must be realistically located. Once the nests placed by streams were discovered, few places were left on BCI for the wrens to go other than the trails, which were heavily used by coatis and other predators. On these trails, new nests were discovered even more rapidly than the original ones. Hence, on Barro Colorado, the only alternative survival ploy was unsuccessful.

This experience draws attention to the fact that while island biogeographic theory may predict overall numbers of species that may become extinct, it does not reveal the underlying causes of extinction for any one species. These causes can be identified by reintroduction designed to test hypotheses, however, and in fact they must be identified before management-level reintroduction is likely to be successful. The fate of the song wrens also illustrates the importance of learned behavior, which for many species probably cannot be maintained in captive populations whose progeny might be used for reintroduction to the wild. This last point must be considered as important as genetic management of small captive populations.

In fact, the wild birds I released had learned a number of behavioral traits in addition to the placing of their camouflaged nests in an appropriate setting. Their foraging behavior, for example, was at least partly learned and would almost certainly be lost in captivity. Song wrens forage close together in leaf litter on the forest floor, so that when one bird flushes a cockroach or katydid from a leaf it has just looked under, the adjacent birds can capture it. This way they flush insects to one another in a group operation. How could one maintain this habit in captive animals that are fed a controlled diet from a food pan? In general, the comparison of the behavior of wild and captive-raised animals following reintroduction offers unique opportunities to learn more about captive management techniques and also about the biology of the animals. How to "keep an animal wild in captivity" is a little-studied and enormously complicated question that will benefit greatly from well-planned reintroduction.

### Acknowledgements

I am very grateful to the Wildlife Preservation Trust International and to the World Wildlife Fund-US for financial support. The staff of the Smithsonian Tropical Research Institute were helpful in many ways.

### References

Karr, J. R. (1982). Avian extinctions on Barro Colorado Island, Panama: a reassessment. *The American Naturalist*, **119**, 20–39.

Long, J. L. (1981). *Introduced birds of the world: The worldwide history, distribution, and influence of birds introduced to new environments*. New York: Universe Books.

Morton, E. S. (1978). Reintroducing recently extirpated birds into a tropical forest preserve. In *Endangered birds, management techniques for preserving threatened species*, ed. S. A. Temple, pp. 379–86. Madison: The University of Wisconsin Press.

Willis, E. O. (1974a). Populations and local extinctions of birds on Barro Colorado Island, Panama. *Smithsonian Contributions to Zoology*, **291**, 1–31.

Willis, E. O. (1974b). Populations and local extinctions of birds on Barro Colorado Island, Panama, *Ecological Monographs*, **44**, 153–69.

Willis, E. O. & Eisenmann, E. (1979). A revised list of birds of Barro Colorado Island, Panama. *Smithsonian Contributions to Zoology*, **291**, 1–31.

*Katherine L. Gross*

Botany Department and Graduate Program in Environmental Biology, Ohio State
University

# 12 Mechanisms of colonization and species persistence in plant communities

One of the distinctive challenges of ecological restoration, as opposed to less ambitious forms of land reclamation, is the creation of communities with the relatively high level of species diversity that typifies many natural communities. Meeting this challenge will depend on the level of our understanding of the factors that determine diversity in natural communities. As Bradshaw points out in Chapter 2, the "acid test" of our understanding of natural communities will be our ability to restore or rebuild a reasonable facsimile of a community that has been damaged or destroyed. This involves not only recreating the structure of the community (does it look the same?), but also reestablishing the crucial functional relationships in the community (do the species interact in similar ways?).

Diversity is a conspicuous feature of many plant communities, and many studies have documented patterns of diversity in plant communities (see discussions in Grime, 1979; Grubb, 1977). This work has been principally descriptive and correlative, but it has directed attention toward a more mechanistic, population-oriented approach of identifying the factors influencing the development and persistence of diversity in plant communities.

There is a growing view among ecologists that plant communities are dynamic assemblages of species, in which all species should be viewed as potential colonists. All plant communities are subject to periodic or occasional disturbances that remove some species and allow others to establish for a time (for a review, see Pickett & White, 1985). Thus, understanding how communities develop following a disturbance – which is the process of succession – is fundamental to developing an understanding of the maintenance, and restoration, of diversity in plant communities. Restoration projects, if they are appropriately designed, can provide ecologists with unique opportunities to test and evaluate competing hypotheses regarding the mechanisms determining the development and maintenance of diversity in plant communities.

For example, Connell & Slatyer (1977) proposed that three principal mechanisms determine the process of succession: facilitation, tolerance, and inhibition. Facilitation occurs when early colonizing species create conditions that favor (or facilitate) the establishment of other species that cannot occur in their absence. The tolerance and inhibition models are alternative mechanisms by which later-colonizing species persist in the community, either by their ability to tolerate the competitive environment or by their ability to inhibit the establishment of other species. At present, there is very little experimental evidence to evaluate the conditions under which each of these mechanisms will operate. It seems likely that these are not mutually exclusive processes, and that they all play some role in succession and maintenance of diversity. Much still remains to be determined, however, about the conditions under which each of these processes will predominate, and about the precise mechanisms underlying each of them (see, for example, Ashby's comments on mechanisms of forest succession in Chapter 7).

These are obviously questions of great importance, not only for restorationists trying to reestablish and maintain diverse communities, but also for ecologists concerned with understanding them. While we have accumulated a considerable amount of information on the patterns of succession and diversity in plant communities, there have been few experimental studies documenting the mechanisms that determine these processes (Connell & Slatyer 1977). The time has clearly come, therefore, to conduct experiments designed to isolate and characterize the factors determining diversity in various kinds of communities (Diamond & Case, 1986). Where possible, such experiments should be conducted in natural communities, but of special interest here is the relationship that might be developed between experimental ecology in natural successional communities and ecologically oriented restoration projects.

In this chapter, I will describe a series of experiments that represent examples of an experimental and synthetic approach to ecological research in successional plant communities. These experiments were designed to answer questions about the factors responsible for the presence (and also the absence) of various plant species in successional old-field communities in south-western Michigan. Although the experiments were designed to answer basic questions about the nature of succession in these communities and were not part of any attempt actually to restore the community in the usual sense, they did involve an attempt to reassemble – or, more precisely, partially reassemble – components of the community. From the standpoint of restoration ecology, therefore, they represent a halfway point between synthetic experiments carried out under rigidly controlled conditions in the laboratory (Section III) and full-scale restoration of entire communities carried out in the field (Section II).

In these experiments I focused my attention on only a handful of species selected from a community composed of perhaps 100 or more species. These species were selected, however, because they had strikingly different life history traits, representative of those of many of the species that typically occur at different times in successional communities. The assumption under-lying this approach is that an understanding of the processes that affect the abundances of these particular species can be extended to the whole community. Although a restorationist probably would not have selected these particular species (which are generally regarded as "weeds"), the results can be applied to other systems. Moreover, the experimental design and procedures described can easily be applied to studies of other species and communities, including those more likely to be of interest to restorationists. My purpose here is to demonstrate in a general way how such a synthetic approach can provide a basic understanding of the mechanisms underlying patterns of community change.

### Mechanisms of succession

Ecologists have long noted that the life histories of the species dominating at various stages during succession tend to follow a predictable pattern (see, for example, Keever, 1950; Egler, 1954; and references in Golley, 1977). Early colonists in a successional sere are typically short-lived, fast-growing species that reproduce early. Older successional fields are generally dominated by longer-lived perennial species, many of which have the ability to reproduce vegetatively (clonal growth). The persistence and abundance of these species later in succession may be due to their competitive ability, their ability to inhibit the growth of other species, or simply their longevity (see Connell & Slatyer, 1977). This shift in dominance from annuals to short-lived, and finally to long-lived, perennials accounts, in part, for the decrease in the rate of species turnover in later successional communities. However, the process of colonization and replacement of individuals and species continues in a later-successional community, although at a much reduced rate.

The pattern of species turnover in successional communities has been documented in a number of classic studies (see Golley, 1977). Purely descriptive studies do not reveal, however, why these predictable changes in species composition occur and what factors are important in determining these patterns. Are they repeatable? Can they be generalized? It was in an attempt to answer these questions that I initiated and carried out the experiments described below.

### Colonization and persistence of species

In south-western Michigan, recently abandoned agricultural fields are usually dominated by common weedy annuals such as ragweed (*Ambrosia artemisiifolia*), lamb's quarters (*Chenopodium album*), pigweed (*Amaranthus retroflexus*) and various smartweeds (*Polygonum* spp). By the second year, these annuals are generally replaced by biennials, short-lived perennials and grasses (J. Cantlon, unpublished; Gross, 1980; Miller, 1985). These species, in turn, are slowly replaced over a period of 10–30 years by longer-lived perennial grasses and forbs. Late-successional (15–40-year-old) fields in this area have a characteristic flora that is dominated by perennial, clonal grasses such as quackgrass (*Agropyron repens*), brome grass (*Bromus inermis*), and timothy (*Phleum pratense*); and a variety of perennial forbs, which typically include several species of goldenrods (*Solidago* spp), hawkweeds (*Hieracium* spp) and several legume species (e.g., *Trifolium* and *Desmodium* spp) (Stergios, 1976; Werner, 1977; Gross & Werner, 1982). Colonization by woody species is slow, apparently because of the absence of seed dispersers, and also because tree seedlings can establish only in sparsely vegetated areas of these fields, where competition from herbaceous dicots is low (Harrison & Werner, 1984; Werner & Harbeck, 1982).

Despite the general shift in dominance over time in these fields from annual to perennial species, species with a biennial or monocarpic perennial life history occur at all stages of succession in these fields. (The distinction between these two life cycles is not important for this discussion, so I will use the simpler term "biennial" here to refer to both.) There is, however, a predictable shift in the abundance of several species of biennials over time. For example, mullein (*Verbascum thapsus*) and evening primrose (*Oenothera biennis*) occur only in recently abandoned fields or in highly disturbed areas in older fields. In south-western Michigan, neither of these species typically persist on a site for more than one generation (Gross, 1980; Gross & Werner, 1982). In later successional fields, two other biennials, Queen Anne's lace (*Daucus carota*) and goat's beard (*Tragopogon dubius*), are common. Unlike mullein and evening primrose, which are limited to newly abandoned, highly disturbed or open fields, Queen Anne's lace and goat's beard occur in relatively stable late successional fields. Both species (especially goat's beard) are commonly found in fields that have been abandoned for as long as 35 years (Gross & Werner, 1982).

The sequence in which these four species appear in successional fields is correlated with differences in their life history traits. Mullein and evening primrose, the two early colonists, both produce a large number of very small seeds that lack adaptations for long-range dispersal. However, seeds of these species can remain viable in the soil for up to 80 to 100 years (Kivilaan & Bandurski, 1981). In contrast, the seeds of the later-appearing Queen Anne's lace and goat's beard are much larger but are produced in smaller numbers. They do not remain viable in the soil, but are adapted for dispersal over long distances, goat's beard by wind (K. L. Gross, unpublished data), and Queen Anne's lace by wind and animals (Lacey, 1981).

While these differences in life history traits suggest reasons for the pattern in which these species appear in succession, simple correlations do not necessarily indicate the reasons for this pattern. It was to determine the mechanisms underlying this pattern that I designed and carried out the following experiments.

### Experimental introductions of species

The experiments I will describe in this chapter were carried out in both the field and the greenhouse. All of them involved bringing together species to test specific ideas about the factors underlying the changes occurring in successional communities over time, and to identify the conditions that promote the establishment and the persistence of species within successional communities. This approach therefore resembles the process of restoration itself, and the results are likely to be useful for the refinement of restoration techniques.

The first series of experiments was carried out in an attempt to determine the factors that influence the time of appearance of these two groups of biennial species: early colonizers like mullein and evening primrose, which typically do not persist in the community; and later colonizers like Queen Anne's lace and goat's beard, which appear later in succession and can persist in the community for many years. Later studies focused on revealing the actual mechanisms responsible for these patterns.

One explanation for the observed temporal pattern of species abundance of these biennials is that their seedlings require different conditions in order to become established. It seems likely, for example, that the early-colonizing species, with their small seeds, might be unable to establish in vegetated sites. In south-western Michigan a dense cover of vegetation becomes established fairly rapidly in successional fields, and species with small seeds might be unable to establish, and thus are rapidly excluded from the community. The later colonizing biennials that I studied may be able to persist in more densely vegetated successional fields, because their larger seeds produce larger seedlings, and are subsequently better able to survive despite the severe competition characteristic of later successional sites.

To determine whether these really are the factors underlying the pattern of appearance of species in these communities, I began a series of experiments in 1978 that involved introducing seeds of these species into different environments and determining their ability to establish in different types of cover at each site. In the first experiments, I sowed a known number of seeds of each species into plots in three old fields that had been abandoned for various lengths of time (one, five and fifteen years). Each of these fields included sites that were essentially bare ground, sites that were heavily vegetated, and sites that were transitional or lightly vegetated. However, the availability of these different types of ground cover differed in the fields – bare ground, in particular, was extremely rare in the older fields.

What I found was exactly what might have been expected on the basis of the seed weights of the four species: seedlings of all four species were more likely to emerge in bare soil, but only the larger-seeded species, goat's beard and Queen Anne's lace survived beyond the seedling stage on the heavily and lightly vegetated sites. These species also survived and grew well on open sites in the youngest field, where they do not naturally appear. This demonstrates that the delayed occurrence of these species in successional fields is not because they cannot establish earlier (e.g., facilitation model rejected). A more likely explanation is the absence of viable (or sufficient) seeds brought in by wind, and their absence in the soil seedbank of a newly abandoned field (Gross & Werner, 1982; Gross, 1980).

It seems, therefore, that the time of appearance of these species in old-field succession depends on the arrival of seed, and that the key factor in their ability to persist in the community is their seed size. Seedlings from species with large seeds are able to survive and to reach maturity in increasingly dense vegetation. However, these four species exhibit a number of other differences that might be as, or even more, important than the seed weight and dispersal traits I have emphasized above in determining their abundance and persistence in successional fields. For example, the morphology or growth form of a seedling can affect the type of ground cover in which it can establish. Seedlings with long, narrow leaves and an upright or vertical growth form, such as goat's beard and Queen Anne's lace, are more likely to emerge through a cover of litter or other vegetation than those that are low-growing or have broader leaves, like mullein and evening primrose. Because seed weight and seedling morphology are correlated in these four species, I can't distinguish which factor accounts for the pattern I observed in the field experiments.

I subsequently tested these ideas by sowing seeds of six species – teasel (*Dipsacus sylvestris*) and burdock (*Arctium minus*), in addition to the four already mentioned – into soil with four different types of cover in a greenhouse. Teasel and burdock both have broad leaves and a relatively low, horizontal growth form as seedlings, but have large seeds similar in weight to those of Queen Anne's lace and goat's beard, respectively. The four cover types were bare soil, litter, an established cover of Kentucky blue grass (*Poa pratensis*), and grass plus litter. The results (Gross, 1984) confirmed the hypothesis that the competitive ability of these species, as measured by seedling growth in the grass cover, is directly related to the weight of their seeds. However, when grown on bare soil the seedlings of the smaller-seeded species grew much more rapidly than those from larger-seeded species. This suggests that on bare soil or in open areas, species with small seeds might have a competitive advantage over larger-seeded species.

The actual cause of these differences in growth rate wasn't determined in these experiments. However, some interesting patterns emerged which I am currently examining in more detail.

Seedlings of these six species allocated biomass very differently: the large-seeded species allocated more biomass to roots, whereas the small-seeded species favored shoots (principally leaves). While precise mechanisms underlying these differences in allocation patterns are not clear, the ecological implications are: species that can differentially allocate biomass to harvest resources most limiting to seedling establishment on a given site will be more likely to persist there.

## Disturbance dynamics and colonization

The experiments I have described here have all focused on how a single factor, seed weight, influences the ability of these biennial species to colonize and to persist in successional communities. The general conclusion from these studies would be that a quite simple and easily identified life history characteristic, such as seed weight or a correlated characteristic, may be primarily what influences the successional status of a species.

On the basis of the experiments described above, one would expect that the seed weight of the species present in successional communities would tend to increase over time. In fact, such a trend has been reported by several people (e.g., Salisbury, 1942; Baker, 1972; Harper, Lovell & Moore, 1970; van der Pijl, 1984). Taken alone, this suggests that there might be a decline in species diversity over time within such communities, since an entire class of plants – those with small seeds – would be excluded from the community. This, however, is not the observed trend. Species diversity generally increases over time in successional communities (Whittaker, 1975; Golley, 1977), and many species with small seeds do persist within late-successional communities (Salisbury, 1942). This suggests that, important as they may be under certain conditions, seed size and dispersal ability are certainly not the *only* factors that determine the establishment of a species in succession. In fact, there are a number of other factors that might account for the observed diversity of species in plant communities (Grubb, 1977, 1986; Grime, 1979). One of these (proposed by Grubb, 1977; 1986) is that some species in late-successional communities can establish and grow in the open sites that occur within the matrix of established perennials in the community. These openings are caused by variation in the growth and senescence of the established perennials and by the activities of burrowing and grazing animals, which create small soil disturbances. While ecologists generally acknowledge the importance of disturbance for the maintenance of diversity in plant communities (Connell, 1978; Pickett, 1980; Pickett & White, 1985), little is known about how such disturbances actually influence the dynamics of communities. It is not clear, for example, how the diversity of the ground cover or the nature and frequency of disturbances in a community affect the diversity of the species that occur there. We do know that plants rapidly colonize these disturbances or open sites in the vegetation. In some cases, the species composition and pattern of colonization on these sites is relatively predictable (Platt, 1975; Platt & Weis, 1977; 1985). In others, the pattern of seedling recruitment varies considerably from year to year (Shmida & Ellner, 1985; Grubb, 1986). This then raises the question of the importance of such events in determining both the diversity of a plant community and the abundance of particular species in that community.

Experiments involving the restoration of communities, or the experimental manipulation of communities in an attempt to recreate features or events that occur naturally, offer excellent opportunities to answer these questions. For the past four years, I have been conducting experiments to determine how the size and time of formation of small disturbances in an old field affect the species that colonize these sites. To do this, I have created local disturbances in a late-successional old field (uncultivated since 1945) by digging up and turning over small patches of soil. These artificially disturbed areas vary from 15 cm to 1 m in diameter and mimic those created in this community by various animals, notably ants, mice, ground squirrels, and gophers. In conjunction with these experimental studies, since 1982 Dr. Deborah Goldberg (from the University of Michigan) and I have regularly surveyed marked plots in three old fields of different successional ages to determine the rate of formation, the size, and the rate of revegetation of areas disturbed naturally by animals.

These observational and experimental studies have led to a number of insights into how localized disturbances affect the species diversity of late-successional plant communities. Our observations indicate, for example, that disturbances caused by animals tend to be very small (80 per cent are less than 20 cm in diameter) and to occur more frequently in areas of the field with sandy soils and sparse vegetation. The size, location, and time of formation of a disturbance all influence the species colonizing the site. For example, in mesic areas of these fields the disturbed areas tend to be larger, but are extremely rare and are rapidly filled in by the growth of vegetative shoots from surrounding plants. Consequently, these sites, though favorable for seedling emergence, rarely serve as colonization sites for seedlings. In contrast, seedlings frequently colonize disturbed sites in drier parts of the field, and it is here that experiments involving artificial disturbances have helped us to understand the factors influencing this process.

My own experimental studies clearly indicate that the species that colonize a disturbed site vary depending on the time of year the disturbance occurs. The question, however, is whether we understand what is going on here well enough to use this information to manipulate the community in a predictable way. To what extent, for example, can we write a prescription for an increase in a particular species in a community simply by specifying the size, timing and distribution of artificial disturbances within the community? My results suggest that it may indeed be possible to do this in certain instances, but that to do so it is necessary to understand the relationship between the critical variable and the response of the vegetation in some detail.

The time of disturbance, for example, may be critical in some years and of relatively little importance in others. For example, the main colonists on disturbed sites created in well-drained, upslope areas of these fields are primarily species whose seeds are present in the seed pool. The most abundant colonists are two native annual grasses, fall panicum (*Panicum dichotomiflorum*) and witch grass (*P. capillare*). Both of these have long-lived seeds and occur only in disturbances, not in the undisturbed vegetation. Though both species produce more seed in larger disturbances, both can regenerate in the relatively small disturbed areas (15–20 cm diameter) common in these fields. In my experiments, the time at which a disturbance was created strongly affected the relative recruitment success of these two species in one year, but had less effect in subsequent years. In the first year, disturbances created in early June were colonized almost exclusively by fall panicum, and those created a month later were colonized almost exclusively by witch grass. This suggests that a manager could regulate populations of these species in a community, favoring one over the other, for example, simply by introducing disturbances at different times during the year. It is important to recognize, however, that this relationship varies from year to year, probably in response to the amount and timing of rainfall. What this means, of course, is that while we have identified a rough relationship between the time of year of a disturbance and a species' ability to colonize it, we have not yet identified the critical factors underlying this relationship. It might, however, be possible to identify these by setting up experiments designed to test inferences based on our observations. The field data suggest, for example, that dry or hot conditions immediately following seedling emergence favor witch grass over fall panicum – a hypothesis that could easily be tested by experiments involving watering natural and artificial disturbances created at different times of the year.

For other species in the community, both the time and the size of disturbance strongly influence their colonization success. For example, although hawkweed (*Hieracium* spp.) seedlings are most abundant in smaller openings, only seedlings that emerge in large openings (1 m) survive to the second year. Furthermore, in these experiments hawkweed seedlings only appeared on disturbances that were created in early June, just prior to the time of seed dispersal, suggesting that hawkweed seeds do not persist in the soil seedbank (see also Stergios, 1976). This, together with the requirement for large openings for successful seedling establishment, which are exceedingly rare in the field, would severely limit the ability of hawkweeds to establish from seed in these communities.

The above findings suggest that hawkweeds would be relatively uncommon in late successional fields when, in fact, they frequently predominate on such sites. They are actually the biomass and numerically dominant species in the drier, upslope areas of the field in which these experiments were conducted. This raises an interesting question as to how these species manage to persist in such fields.

To answer this question we must recognize that there are alternative means by which an established plant population can persist in a community. I have found, for example, that *Hieracium* plants are common on both natural and artificial disturbed sites one to two years after they are created. However, generally the plants colonizing the disturbances are not from seedlings, but from vegetative shoots produced by plants surrounding the disturbance. This suggests that the persistence of hawkweeds in old fields is primarily due to vegetative growth rather than to seedling recruitment. Similar patterns have been found for clonal perennial species in other communities, notably *H. floribundum* in old fields and pastures in Canada (Thomas & Dale, 1975), and two species of goldenrod (*Solidago canadensis* and *S. juncea*) in old fields in Michigan (Goldberg & Werner, 1983). Like our hawkweeds, these species establish as seedlings only in relatively large openings, and apparently represent a group of plants that become established early in a successional sere, when large patches of bare soil are available. These species persist in older fields because of clonal growth.

### Mechanisms of persistence

These observations and experiments have led me to recognize three broad categories of plants reflecting distinctive life history patterns and behavior in old fields and similar communities (Table 12.1). This classification scheme differs in several respects from those suggested by other authors (e.g., *r*- and *K*-selected species (MacArthur & Wilson, 1967), and the "ruderal", "stress-tolerant" and "competitor" scheme of Grime, (1977) because it focuses on the alternative means by which species may persist *within* the same community. The *regenerators* (which include species such as goat's beard and Queen Anne's lace) are short-lived, nonclonal species that produce large seeds, and can persist in late successional (or similar) communities because their seedlings can tolerate high levels of

Table 12.1. *Three alternative mechanisms of persistence in plant communities*

| Characteristic | Regenerators | Fugitives | Persistors |
|---|---|---|---|
| Seed weight | Large | Small | Small |
| Seed number | Few | Many | Many |
| Dispersal distance | Far | Near | Far |
| Seed longevity | None | Long-lived | None |
| Seedling "safe site" | Many | Bare soil | Bare soil |
| Adult longevity | Intermediate, non-clonal | Short-lived | Long-lived, clonal |
| Persistent stage | Seedlings and adults | Seeds | Adults |

competition. The *fugitives* (for example, mullein, evening primrose and the two *Panicum* species) are short-lived plants that have small, long-lived seeds. Interestingly, *persistors* resemble fugitives in many characteristics, except that their seeds are not long-lived in the soil and they require relatively large disturbed areas for seedlings to establish. These species are able to persist and even to increase in abundance in late-successional and/or densely vegetated communities because they are perennials and capable of extensive vegetative propagation. Examples of persistors are the hawkweeds and goldenrods described above.

The value of this classification scheme is that it focuses attention on what I feel to be the critical factors underlying the ecological behavior of these species in late-successional communities. It points out both similarities and differences that are important but that might easily be overlooked in determining both the temporal distribution and abundance of different plant species in successional communities (Table 12.1). Although I have included as examples here species that are typically found in old fields, I am confident the scheme can be applied to many other plant communities. For example, this scheme has obvious parallels with the categories devised by Blewett & Cottam (see Ch. 17) to describe the changes in abundance of plants on a restored prairie over a 25-year period. Their "increasers" are most likely regenerators or persisters. Their "decreasers" and "sensitive" species might be regenerators or, more probably, fugitive species whose "safe site" has decreased in abundance. Their "no change" species are clearly persisters.

The value of the classification scheme I have proposed above is that it provides some explanation as to *why* these species behave as they do – information that is potentially very important in devising appropriate management or restoration plans for a community. It would seem the appropriate next step would be to compare the life history traits of the "preferred" prairie species in these restored communities to those I have listed in the above classification scheme, and to design experiments that would be predicted to result in an increase in their abundance. The point here is that, in general, experimental studies involving both the disruption of communities and attempts to manipulate them in precise ways do seem to offer ways both of sharpening the definition of categories such as those I've proposed and of identifying the mechanisms that account for the patterns of behavior to which they refer.

### Conclusion

The ideas and hypotheses I have presented regarding the importance of regenerative traits in determining the diversity and abundance of species in plant communities cannot provide, in themselves, a prescription for restoring a damaged or destroyed community. There is clearly much room for further experimental work to address the issues and questions discussed here and described in other chapters in this volume. The field of restoration ecology, as outlined in this volume, can provide new opportunities to test and to modify these hypotheses and to determine the limits and the extent of their applicability in a variety of systems. The processes I have emphasized in my research may not operate, or may play quite a small role in determining the dynamics of species in restored communities, especially those in which the soil structure has been badly damaged. For example, elsewhere in this volume, both Bradshaw and Miller describe the processes regulating the recovery of communities that have undergone more profound disruption than is typically caused by agriculture. Under such severe conditions, abiotic factors such as the structure of the soil, the availability of specific nutrients, and mycorrhizal associations may be of overriding importance in determining the time, course and pattern of succession and recovery. Of interest here, however, is the clear demonstration that, at least under certain conditions, interactions between species, like those influencing community development in highly simplified experimental systems, such as Gilpin's laboratory fruit fly communities (Ch. 10), do influence community change in "real" communities in the field. What remains is to work out, for each situation, the complex of factors governing changes in species composition and to determine what relation (if any) this has to the life history traits, morphology, and physiology of the species involved. If the population processes I have focused on here can be shown to operate in a similar way in a diverse array of successional or restored communities, then we will have moved considerably closer to an understanding of the processes determining patterns of species' presence and abundance in plant communities generally.

However, such an understanding will emerge from restoration projects *only* if they incorporate properly designed experiments with appropriate controls and sufficient replication to answer specific questions. This argues strongly for the incorporation into restoration projects of experimental studies designed to test specific hypotheses. More descriptive studies or restoration projects that result from a hit-or-miss tinkering (successful or not) will do little to advance our understanding of how natural plant communities function. There is currently much debate in community ecology as to what processes account for the observed diversity of plants and animals (see review in Diamond & Case, 1986). There is clearly a need for pluralism in approaches to the study of communities, and restoration ecology may provide an ideal way of demonstrating how pluralistic approaches can successfully bridge the gaps in our understanding of natural and managed communities. In this way it can provide valuable information on both the processes shaping natural communities and what needs to be done to preserve, protect or restore them.

### Acknowledgements

Support for this work was provided by funds from the National Science Foundation (BSR 83-14742) and the Ohio State University. The comments on earlier drafts of this chapter by Gary Mittelbach, Deborah Goldberg, Carolyn Wilczynski, Alice Winn, Tom Miller and Bill Jordan improved its content and clarity.

### References

Baker, H. G. (1972). Seed weight in relation to environmental conditions in California. *Ecology*, **53**, 997–1010.

Connell, J. H. (1978). Diversity in tropical rainforests and coral reefs. *Science*, **199**, 1302–10.

Connell, J. H. & Slatyer, R. (1977). Mechanisms of succession in natural communities and their role in community stability and organization. *The American Naturalist*, **111**, 1119–44.

Diamond, J. & Case, T. J. (eds) (1986). *Community Ecology*. New York: Harper & Row.

Egler, F. E. (1954). Vegetation science concepts. I. Initial floristic composition – a factor in old-field vegetation development. *Vegetatio*, **4**, 412–7.

Goldberg, D. E. & Werner, P. A. (1983). The effects of size of opening in vegetation and litter cover on seedling establishment of goldenrods (*Solidago* spp). *Oecologia* (Berlin), **60**, 149–55.

Golley, F. B. (ed.) (1977). *Ecological Succession*. Stroudsburg: Dowden, Hutchinson & Ross.

Grime, J. P. (1977). Evidence for three primary strategies in plants and its relevance to ecological and evolutionary theory. *The American Naturalist*, 111, 1169–94.

Grime, J. P. (1979). *Plant strategies and vegetation processes*. New York: John Wiley.

Gross, K. L. (1980). Colonization by *Verbascum thapsus* (Mullein) in an old field in Michigan: experiments on the effects of vegetation. *Journal of Ecology*, 68, 919–27.

Gross, K. L. (1984). Effects of seed size and growth form on seedling establishment of six monocarpic perennial plants. *Journal of Ecology*, 72, 369–87.

Gross, K. L. & Werner, P. A. (1982). Colonizing abilities of "biennial" plant species in relation to ground cover: implications for their distribution in a successional sere. *Ecology*, 63, 921–31.

Grubb, P. J. (1977). The maintenance of species richness in plant communities: the importance of the regeneration niche. *Biological Reviews*, 52, 247–70.

Grubb, P. J. (1986). Problems posed by sparse and patchily distributed species in species–rich plant communities. In *Community Ecology*, ed. J. Diamond & T. J. Case, pp. 207–25. New York: Harper & Row.

Harper, J. L., Lovell, P. H. & Moore, K. G. (1970). The shapes and sizes of seeds. *Annual Review of Ecology and Systematics*, 1, 327–56.

Harrison, J. S. & Werner, P. A. (1984). Colonization by oak seedlings into a heterogeneous successional habitat. *Canadian Journal of Botany*, 62, 559–63.

Keever, C. (1950). Causes of succession on old fields of the Piedmont, North Carolina. *Ecological Monographs*, 20, 229–50.

Kivilaan, A. & Bandurski, R. S. (1981). The one hundred–year period for Dr Beal's seed viability experiment. *American Journal of Botany*, 68, 1290–2.

Lacey, E. P. (1981). Seed dispersal in wild carrot (*Daucus carota*). *Michigan Botanist*, 20, 15–20.

MacArthur, R. H. & Wilson, E. O. (1967). *The Theory of Island Biogeography*. New Jersey: Princeton University Press.

Miller, T. E. (1985). Competition and complex interactions among species: community structure in an early old–field plant community. PhD dissertation, East Lansing: Michigan State University.

Pickett, S. T. A. (1980). Non-equilibrium coexistence of plants. *Bulletin of the Torrey Botanical Club*, 107, 238–48.

Pickett, S. T. A. & White, P. A. (1985). *The Ecology of Natural Disturbance and Patch Dynamics*. Orlando: Academic Press.

Platt, W. J. (1975). The colonization and formation of equilibrium plant species associations on badger disturbances in a tall–grass prairie. *Ecological Monographs*, 45, 285–305.

Platt, W. J. & Weis, I, M. (1977). Resource partitioning and competition within a guild of fugitive prairie plants. *The American Naturalist*, 111, 479–513.

Platt, W. J. & Weis. I. M. (1985). An experimental study of competition among fugitive prairie plants. *Ecology*, 66, 708–20.

Salisbury, E. J. (1942). *The Reproductive Capacity of Plants*. London: G. Bell & Sons.

Shmida, A. & Ellner, S. P. (1985). Coexistence of plant species with similar niches. *Vegetatio*, 58, 29–55.

Stergios, B. G. (1976). Achene production, dispersal, seed germination and seedling establishment of *Hieracium aurantiacum* in an abandoned field community. *Canadian Journal of Botany*, 54, 1189–97.

Thomas, A. G. & Dale, H. M. (1975). The role of seed reproduction in the dynamics of established populations of *Hieracium floribundum* and a comparison with that of vegetative reproduction. *Canadian Journal of Botany*, 53, 3022–31.

van der Pijl, L. (1984). *Principles of Dispersal in Higher Plants.* 3rd edn. Berlin: Springer–Verlag.

Werner, P. A. (1977). Colonization success of a "biennial" plant species: experimental field studies of species cohabitation and replacement. *Ecology,* 58, 840–9.

Werner, P. A. & Harbeck, A. L. (1982). The pattern of tree seedling establishment relative to staghorn sumac cover in Michigan old fields. *The American Midland Naturalist,* 108, 124–32.

Whittaker, R. H. (1975). *Communities and Ecosystems.* 2nd edn. New York: Mac-Millan.

*Michael L. Rosenzweig*

Department of Ecology and Evolutionary Biology, University of Arizona

# 13 | Restoration ecology: a tool to study population interactions?

The community ecologist studies sets of populations that interact, on a pairwise basis, in three fundamentally different ways: through predation; competition; and mutualism. These systems can be described by complex networks of dynamical equations, and these abstract representations reveal, as nothing else can, the close conceptual relationship between ecological systems and systems in electronics, biochemistry, astronomy, economics, atmospheric physics and many other systems being studied by scientists.

In modeling a system with such equations, the scientist focuses attention on the so-called state variables. In ecology these are usually population sizes; in biochemistry they may be the concentrations of reacting substances, and in economics the amounts of capital available. For such models to be useful, however, all of the state variables must be measurable and must be represented in the model. The model then consists of equations that describe the change (or motion) of the state variables through time – that is the "trajectory" of the system. The most important features of these equations are the interactions they define between the different state variables. And if we are able to write equations that predict the size of the system's state variables at any time, we can then claim (in some formal sense) that we understand the system.

To visualize this, imagine our system as two pendulums connected by a spring (Fig. 13.1). This simple physical system is analogous to an ecological system comprising a predator and its prey. The relevant state variables are the positions of the two pendulum masses which are analogous, respectively, to the size of the predator and prey populations. What the scientist (whether ecologist or physicist) wants to know is the natural behavior of the system, the parameters that govern the motions of the pendulums and the strength of the interaction between them. That is, having specified the structure of the model, one wants to know exactly how a change in one state variable will affect the other.

A favorite technique for accomplishing such a measurement is called perturbation analysis. The basic idea behind it is quite simple: apply some external force to a system to move it off track (or at least away from an equilibrium point), then watch it carefully as it returns to its natural state. The variables at work in the system can then be inferred by measuring the rates of return as the scientist pushes the system away from its natural state. In the pendulum system, for example, one pulls back one mass and then observes how soon and to what degree its oscillation is transferred to the second mass. In ecology, one might bring about an analogous perturbation by removing half of a prey population, leaving the predator population intact.

Biochemists and other laboratory scientists apply perturbation analysis all the time. For them it is relatively easy and certainly fast. If, for example, they need to explore the interaction coefficient of an enzyme and its substrate, they set up a carefully controlled environment, allow the compounds to

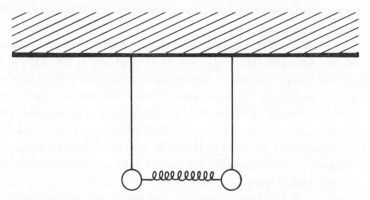

*Fig. 13.1.* A simple physical system – two pendulums linked by a spring – models the ecological relationship between predator and prey.

attain equilibrium, then apply microperturbation. Within a fraction of a second, their system has returned to equilibrium, they have charted its dynamics along the way and they have the data to calculate the coefficients.

Obviously, ecological systems cannot be as carefully controlled as biochemical ones. They also require far more than microseconds to yield their data. Yet ecologists are not quite in the position of astronomers – they can and do perform perturbation work. However, because perturbation analysis is so difficult to perform in ecology, it is important to establish that it is worth doing. I shall proceed by attempting to do that. Then I shall point out some special advantages that restoration ecology offers to perturbation work, advantages that accrue on the one hand because restoration work is viewed as necessarily a large-scale, long-term investment, and on the other because only restoration or other synthetic approaches can deal with the range of state variable alterations that will be necessary to expose community structures.

## Measuring community interactions

Growth of populations of single species is governed by growth rates ($r$) and carrying capacities (K). The added coefficients that govern interactions between species were first defined by Lotka (1925) and Volterra (1926) as terms $\alpha_{ij}$, which describe the effect of a single individual of species $j$ on the growth rate of a population of species $i$. This definition practically demands that we perform perturbation experiments to measure these terms – that is, that we actually add individuals of species $j$ while simultaneously monitoring the change in growth rate of species $i$.

There are, however, problems with this approach. For one thing, as Diamond points out in Chapter 23, it might be illegal, immoral or logistically infeasible actually to modify the state of species $j$. Furthermore, since ecological communities are subject to considerable statistical "noise", any effect of a small change in the density of species $j$ might be obscured. As a result, ecologists have sought other methods to obtain alpha values.

### The overlap approach

Between the 1920s and the 1960s, no one actually tried to measure alpha. Rather, ecologists debated the qualitative features of various alphas, arguing whether they were positive, zero or negative. The entire question of interspecies competition was discussed in terms of whether there were negative alpha values for the interaction between species $i$ and $j$.

All this changed radically with two seminal publications (MacArthur & Levins, 1967; Levins, 1968). Both pointed out that alpha, or the set of alpha values in an entire community, was worthy of quantitative study. Levins's contribution did not address the practical side, the actual field measurement of alpha, but MacArthur & Levins (1967) did focus on this issue. Their idea was to relate alpha to the extent that species overlap in their use of resources. In other words, they recommended first finding out with what frequency each species of a pair utilizes resources of each sort. They then proposed the following rule: if two species utilize virtually the same resources in the same proportions, they probably have a high competitive alpha; smaller overlaps imply proportionately smaller alpha values (Fig. 13.2). They reduced their rule to a mathematical expression, relating overlap to alpha.

Unfortunately, MacArthur & Levins's technique has been discredited. There are a number of logical and operational problems with their approach (Abrams, 1975, 1983; Dayton, 1973; Turelli, 1978). One serious blow emerged from an experiment Schroder & Rosenzweig (1975) performed on two species of kangaroo rats (*Dipodomys*). These species overlap greatly in their use of habitats and seeds, suggesting that the alpha values defining the interaction between them should be close to unity and implying that a change in the density of one species should result in an almost perfectly compensatory change on the part of the other. To test this prediction, Schroder and I performed a set of perturbation experiments. In each we removed, as best we could, one species from the system. Surprisingly, there was no response on the part of the other species. Instead of being close to unity as predicted, the alpha values were actually close to zero. Such results soon discredited the idea of measuring alpha by measuring overlap in resource use.

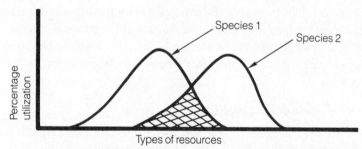

*Fig. 13.2.* Resource utilization curves for two species. The idea was that the amount of overlap (shaded area) would provide an index to the interaction coefficient α between the species.

## The census regression technique

In response to the inadequacy of overlap studies, another method has been suggested. This one, which may be called the census regression technique, was independently suggested by Schoener (1974) and by Pimm (Crowell & Pimm, 1976).

Their idea was to take many independent censuses of the various species in order to determine how changes in their populations were related. If species A tends to be low when B is high and vice versa, then A and B are probably competing. If A and B tend both to be high at the same time, or low at the same time, then they are probably mutualists. On the basis of such censuses, moreover, it is even possible to produce an estimate of alpha for the two species.

Census regression is not nearly so quick and cheap as the overlap approach, yet it promised to outperform perturbation analysis. It was tested by a computer simulation (Hallett & Pimm, 1979), which assumed a community in a homogeneous environment comprised of pairs of species whose carrying capacities differ by no more than one order of magnitude.

Under such circumstances this approach actually works, but of course such circumstances are unrealistic. A major obstacle to applying the census regression technique, recognized immediately by Crowell & Pimm (1976), is that environments are far from homogeneous. Homogenizing them statistically is a job that may be tried in a number of ways, but all are variations on the theme of ascribing the different censuses to habitat differences by using a set of statistical tools: regression analysis.

Recently my colleagues and I (Rosenzweig, Abramsky & Brand, 1984; Rosenzweig *et al.*, 1985) have attempted to estimate alpha values for several species pairs in a community of gerbils, using the technique of census regression. We examined the effects of varying the method of statistical habitat homogenization and discovered that the results depend entirely on the method used. Using some methods, we obtained positive alphas for certain interactions; using others, we got negative alphas for exactly the same interactions. We also got many cases of significant competition using one method, and no significance at all using another.

Now the same data cannot simultaneously reveal a competitive interaction between species and an absence of interaction or, even worse, mutualism. There must be a problem somewhere. At least 32 of 150 alpha estimates we determined from the five rodent species we studied are internally inconsistent and therefore have to be wrong. To our dismay, we can find no *a priori* reason to prefer any of the six methods of habitat homogenization we used.

In another study (Abramsky, Bowers & Rosenzweig, 1986), this one on a pair of Rocky Mountain bumblebee species, we were able to compare the results of the census regression technique with those of actual perturbation experiments. Again, the results were ambiguous. One $(\alpha_{ij})$ was estimated to be virtually zero by both techniques. However, the reverse $(\alpha_{ji})$ was also estimated at zero by regression whereas perturbation showed it to be significantly negative.

Thus, it would seem we really do not yet have a technique to replace direct perturbation analysis for deciphering population interactions. So we must continue to perform perturbations: but how large should they be? and how should they be organized? The answers to these questions are that they should be very large – in fact about as large as restoration projects – and that they could very easily be organized the way restoration projects are. Let us see why.

### Severe nonlinearity in ecosystems

The reason that ecologists should (and do) doubt the general validity of small perturbation analyses is that they have good reason to suspect that the dynamics of ecological systems are highly nonlinear. In practice, this simply means one cannot extrapolate ecological results very much. For example, Norton Juster once playfully wrote that "If a beaver two feet long with a tail a foot and a half long can build a dam twelve feet high and six feet wide in two days, all you need to build Boulder Dam is a beaver sixty-eight feet long with a fifty-one foot tail". But of course we know that a 68-foot-long beaver would die of suffocation, its lungs crushed by its own weight. The trouble is one cannot extrapolate: the nature of a beaver's problems when it is beaver-size is completely irrelevant when it becomes Boulder Dam size.

To understand this better for interacting systems, consider again the spring/pendulum system of Figure 13.1. We conceptualized the study of the strength of the interaction between the pendulums by making small displacements in one mass and observing the effect this had on the other. If, however, we make very large displacements in one mass we might obtain quite different results. If, for instance, we make the change very rapidly, the response might be limited by the viscosity of the surrounding air. Similarly, if we introduce a large perturbation instead of a small one, raising the first mass very high and changing the angle of its linkage to the second mass by a large amount, the transmission of the resulting force to the second mass may change dramatically. These effects are real, and could be quite important if we actually had to assemble the system ourselves.

The interaction of pairs of species is often represented as occurring in a state space defined by the densities of the two species (Fig. 13.3). Figure 13.3 shows that two species have, in principle, an equilibrium at which their densities might remain constant, though they are also subject to a certain amount of temporal variation, reflecting the background noise prevailing in the system. Around their equilibrium point is a zone of small perturbations which, in the absence of noise, would reveal the strength of the species' interactions. Surrounding this is a much broader zone that represents major perturbations, and that includes the extreme perturbations of complete removal or reintroduction of species.

For two reasons, small perturbations may be ineffective in revealing the dynamic ecological relationships in Figure 13.3: the first is that small perturbations may be indistinguishable from background noise; the second is that small perturbations perforce explore the dynamics in only a small region of state space. As the examples of the next two sections will show, one cannot extrapolate results from such a small region to the region of major perturbations. Yet the behavior of the system following major perturbation is what should most concern us.

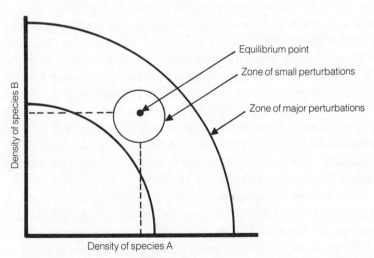

*Fig. 13.3.* A two-dimensional state-space defines the interaction between species A and B.

No one is surprised if an ecosystem can repair a little damage to itself. The real question is, what are the consequences of major damage? Can species associations build back from the brink of catastrophe? And, if so, how? Are the processes involved the same as when the minor damage is repaired? Do they tend to move toward a unique structure? Or does the final structure vary, reflecting the exact state of the association when it begins recuperation? Restoration following major perturbation may be the only way to answer these questions.

### Paradox of enrichment

One case that illustrates the inadequacy of small perturbations involves the interaction ecologists call biological exploitation, or predation. This occurs when one species takes advantage of a second. In dynamic terms, alpha is positive ($\alpha_{ij}$) in one direction and negative in the other ($\alpha_{ji}$).

MacArthur and I (Rosenzweig & MacArthur, 1963) analyzed predation in part by linearizing it in a small region of state space. This method is the precise mathematical equivalent of small perturbation studies. We predicted that predators that are too proficient at catching their victims would send the system into unstable (growing) oscillations that might (and in the laboratory do) result in extinction. However, several investigators (Kolmogorov, 1936; Gilpin, 1972; May, 1972), applying non-linear methods, showed that the system's oscillations are always bounded and by themselves can never lead to extinction. Instead, extinction occurs in such a system only as a result of some accident occurring near the low point of a cycle.

Non-linear predation theory has offered an insight into the dynamics of artificial eutrophication (Rosenzweig, 1971). It revealed a phenomenon called the *paradox of enrichment*. Suppose the death rate of an important prey species is attributable both to predation and to a shortage of resources. Now a well-meaning manager supplies an abundance of the limiting resources. Instead of helping matters, this might actually send predator and victim into an unstable, high amplitude oscillation from which neither will recover. The paradox is that feeding a population that is starving can actually exterminate it, if it is under predatory regulation.

Once noticed, the cause of the paradox is not hard to fathom. A victim population, when close to its carrying capacity, is strongly influenced by intraspecific competition. This is a negative feedback process: as the number of victims increases, the competition becomes more severe and the growth rate of the victim population declines. Such negative feedback processes help to keep the population within bounds. If, however, we now add more of whatever resources had been in short supply, we reduce the competition and its stabilizing effect.

Meanwhile, there is another interaction, a mutualistic one, that unlike competition is not accentuated by the resource enrichment. When a predator feeds successfully it tends to become satiated and captures its prey more slowly. Inasmuch as victims help to satiate predators, their relationship with each other is to some extent mutualistic: the larger the victim population, the smaller the burden of predation on each one. This is a positive feedback process and leads to instability: the more victims, the lower the burden of predation and the faster the victim population grows.

Now the cause of the paradox is evident. Enrichment removed the victims from the stabilizing influence of competition toward the destabilizing one of mutualism.

Linear analysis could never have revealed this paradox because linear analysis is always based on the assumption that the interactions between and within species are constant. Non-linear analysis not only allows them to vary in different regions of the state space, it predicts how they should vary.

The paradox of enrichment may have a great deal to do with the dynamic aberrations that often result from artificial eutrophication in lakes. There, not a kindly human manager, but a rapacious or unthinking human society provides the enrichment (leftover phosphates, etc.). But the population dynamics is blind to the motivation, so the result is the same.

### The ghost of competition past

Behavioral ecologists often find that optimal foraging behavior is characterized by a severely non-linear pattern. As a particular variable is tuned (that is, altered slowly and continuously), the individual it affects does not change its behavior. Then, all of a sudden, the tuning takes the variable across some threshold and the individual alters its behavior greatly. Further tuning in the same direction does not cause further behavioral change.

Anyone who has tinkered with the horizontal hold of an old-fashioned television set has observed this pattern. There is a range of settings, over which there is no change in the picture. Turn past this range, however, and the picture breaks up into chaos.

An example of a severely non-linear behavioral pattern is a forager's choice of patches to utilize, and the way this choice varies in response to changes in population density. Assume, for example, that there are two types of habitats in which the individuals of a species can forage profitably.

One of these, however, is so superior that when the species is rare, individuals of the species reject all opportunities to utilize the less favorable type because they are better off spending their time looking for patches of the superior "primary" type.

Now imagine that the forager's population is growing. Each increment further reduces the quality of the primary patches, because there are more foragers exploiting them. At some point they are no better than the secondary ones, and the primary and secondary patches should be exploited equally. Thus, somewhere between the species' carrying capacity and zero, there must be a census point at which the individuals switch abruptly from selective to opportunistic behavior. They should not switch gradually, but sharply and discretely (see the study of gerbil behavior described in Rosenzweig & Abramsky, 1985). The threshold density at which the change takes place is called an isoleg point.

The resources of patches may also be reduced by other competing species. So the density at which a forager needs to switch from selectivity to opportunism can be expected to depend upon the density of these competing species, too. If we had a graph on which we plotted the densities of, say, two foraging species against each other, there would be many isoleg points, each reflecting a combination of forager densities at which one of the foragers achieves its behavioral threshold. The collection of these isoleg points is a line called simply an isoleg. On one side of the line a species is selective, on the other it is opportunistic (Rosenzweig, 1979, 1981, 1985; Pimm & Rosenzweig, 1981; Rosenzweig & Abramsky, 1986; Brown & Rosenzweig, 1986).

Perhaps the most interesting example of an isoleg graph for two species is the following. There are two patch types, a and b. At any instant there is no perceptible difference between patches of a given type. There are also two foraging species, A and B. An individual of either species can profitably use a patch of either type, but the species have distinct preferences: individuals of A do better in a; individuals of B do better in b. (*Note*: preference is determined when A and B are very rare.) Finally, we assume that the two species actually utilize the same resources when they visit the same patches.

The preceding model suggests the following predictions (which are summarized in Fig. 13.4). The state space should be divided into three regions by two positively sloped isolegs (Rosenzweig, 1981; Brown & Rosenzweig, 1986). These areas should be elongated, with their principal axis running NE–SW. Brown & Rosenzweig (1987) have shown that these isolegs never cross, so there is no other region. The central region should be characterized by complete selectivity on the part of both species. In other words, in this region their habitat utilizations should be completely separate and distinct. In the upper region, B is opportunistic, so it overlaps A in its secondary patch. In the lower region, A is the opportunist, so again there is overlap. One may surmise that there is no competition in the central region. But there should be competition in the flanking ones, because of the overlaps in habitat use.

If the equilibrium point of this interaction lies in the central region, then at equilibrium the two species will appear to be irrelevant to each other. They will use entirely separate patch types and never compete. A small perturbation experiment will confirm this. In fact, it has been shown that

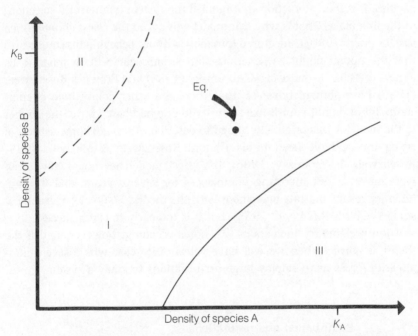

Fig. 13.4.  *The ghost of competition past.* Isolegs divide the state space of two species into three zones. The solid line is A's isoleg; the dashed line is B's. In the central zone (I), there is no niche overlap and no overt evidence of competition. Yet competition is overt in the flanking regions (II and III), and influences the system everywhere. The equilibrium point of the system (Eq) should fall in zone I. But at equilibrium, A's density must be less than its carrying capacity $K_A$ and B's density must be less than $K_B$ because of the competition in zones II and III. This model provides another example of the severe non-linearities that make it necessary for ecologists to carry out large-scale, synthetic experiments in order to understand ecosystems.

the equilibrium point will most likely be in the central region (Pimm & Rosenzweig, 1981; Brown & Rosenzweig, 1986).

Now imagine a major perturbation. This could easily take us across one of the isolegs in Figure 13.4. There we would see one of the species using both patch types. The two species would now affect each other's growth rates negatively, and we would find that they do compete.

Actually, in the total absence of its competitor, each species would have to be opportunistic at its carrying capacity (Morisita, 1950; Fretwell, 1972). Thus, it is the competition that restricts the species' actual behavior at equilibrium. And this is so despite the fact that the competition is invisible to analyses done by small perturbations. The competition lurks like a ghost in the flanking regions, keeping equilibrium densities below carrying capacities. It is this interaction that molded the system (via natural selection) in the first place. That is why this model was called the *Ghost of competition past*. Connell (1980) conjectured (probably without believing his conjecture) that the notion might be generalizable: perhaps competition tends to be almost invisible. Rosenzweig & Abramsky (1986) and Brown & Rosenzweig (1986) have shown, however, that there are other competitive circumstances that do not vanish like a ghost. So one had best restrict the epithet to the model it was originally coined to describe. Although some aspects of isoleg theory have been tested (Pimm, Rosenzweig & Mitchell, 1985; Rosenzweig & Abramsky, 1986), the ghost model has not. I have been working on a test involving bumblebees for several years, and the preliminary results indicate agreement with the theory. So do biogeographical studies of birds (Diamond, 1985), but it is too early to claim success.

If the preliminary indications are not misleading, however, and if the theory is correct, then we will have yet another case where severe non-linearity forces us to employ large perturbations to study a system.

### Perturbation and restoration

Having made the point that large-scale perturbations are necessary to understand the organization and operation of natural ecosystems, we may turn to a practical question. Who will do these experiments? Unfortunately, academic ecologists are rarely able to manipulate at the scales of time and space required. Moreover, wholesale removals of species from ecosystems have been tried often enough. No one should have the heart to do any more – certainly not intentionally.

It should be obvious that restoration ecology is the answer. From an analytical point of view, putting species back into a community is just as good as removing them, and it is far more attractive from every other point of view. Restoration projects (in contrast to projects involving removal of species) should be able to generate the long-term public commitment required; and what community will settle for restoration of a single acre when it might afford to restore a thousand? Restoration projects thus promise to be just the sort of large, long-term perturbation experiments that ecologists need to perform. Adding species to a natural system – if carefully done – can help ecologists get many of the answers they need, though the caveat "if carefully done" is crucial.

Of course, we must supply the right equipment to the right scientists in order to record and analyze the experiments professionally. And – more difficult – we will need to admit that no good experiment gets done without adequate controls and replications. Now we are getting to the crux of the matter.

Controls mean places where we prevent an ecosystem from being restored. Will the people committed to restoration consent to that? Moreover, we will need to limit, perhaps to small sets at a time, the species that are returned to the environment. No scientist can learn much if forced to work simultaneously with many variables, and there is no reason to expect the restoration ecologist to be an exception. We will probably also need to remove species that we have successfully introduced so that we may discover whether the order of introduction is important (see Ch. 10). Other ecologists, such as Kay Gross and Pat Werner who describe their work in Chapters 12 and 22, have done such one-factor-at-a-time synthetic work on a modest scale. But does anyone honestly believe that such an approach will not cause an uproar when pursued on a large scale? The enthusiasm with which the amateur is likely to set about doing ecological restoration may be a curse in disguise for restoration ecology. The mandate likely to be given to the professional ecological restorer to fix everything "like new" may be a juggernaut that reduces the opportunity for carrying out meaningful basic ecological research.

Yet if we truly want to understand what we are doing, both as ecologists and as restorationists, we must sort out causes and effects. The methods of science may sometimes seem cold, but they have repeatedly proved to be the best ways of doing the sorting. Only after we apply them will we find out whether there are patterns and regularities in ecosystems. Only after we apply them can we develop a reliably predictive science, and demonstrate that this whole business is not so capricious that we must construct and study each individual ecosystem to reveal how it functions and how to protect it.

Howard Levene once said that the twentieth century is in for two nasty ecological surprises. The first is what happens when unreasonable densities of human beings begin to acquire extraordinary technological abilities to modify environments: this is already happening. The second happens when humanity turns to ecologists to solve the resulting problems. Restoration ecology, applied with restraint, patience and care, is probably the best way to avoid this second surprise.

No one pretends all this will be easy. It may not even be politically feasible; but it is scientifically possible and it is important to try. An ultimate goal of ecologists is to have a science capable of predicting the structure and function of natural communities and ecosystems, especially those whose component species have had a chance to adapt to whatever perturbation society proposes to impose. We ecologists know that we are very far from being able to do such a thing. We also know that to achieve such a goal will take the fervor and self-effacing dedication of the generations that built the great European Gothic cathedrals. Will having the tools of restoration ecology inspire us to that degree?

### Acknowledgements

Dr M. Gilpin took extraordinary care and spent much effort editing this chapter. He is to be credited entirely with the double pendulum analogy. The United States National Science Foundation (BSR 81–03487) supported the work.

### References

Abrams, P. (1975). Limiting similarity and the form of the competition coefficient. *Theoretical Population Biology*, **8**, 356–75.

Abrams, P. (1983). The theory of limiting similarity. *Annual Review of Ecology and Systematics*, **14**, 359–76.

Abramsky, Z. A., Bowers, M. & Rosenzweig, M. L. (1986). Detecting interspecific competition in the field: testing the regression model of Crowell and Pimm. *Oikos*, **47**, 199–204.

Brown, J. S. & Rosenzweig, M. L. (1986). Habitat selection in slowly regenerating environments. *Journal of Theoretical Biology*, **123**, 151–71.

Connell, J. H. (1980). Diversity and the coevolution of competitors, or the ghost of competition past. *Oikos*, **35**, 131–8.

Crowell, K. L. & Pimm, S. L. (1976). Competition and niche shifts of mice introduced onto small islands. *Oikos*, **27**, 251–8.

Dayton, P. K. (1973). Two cases of resource partitioning in an intertidal community:

making the right prediction for the wrong reason. *American Naturalist*, **107**, 662–70.

Diamond, J. M. (1985). Evolution of ecological segregation in the New Guinea montane avifauna. In *Community Ecology*, ed. J. M. Diamond & T. J. Case, ch. 6. New York: Harper & Row.

Fretwell, S. D. (1972). *Populations in a Seasonal Environment*. Princeton: Princeton University Press.

Gilpin, M. E. (1972). Enriched predator–prey systems: theoretical stability. *Science*, **177**, 902–4.

Hallett, J. G. & Pimm, S. L. (1979). Direct estimation of competition. *American Naturalist*, **113**, 593–600.

Kolmogorov, A. N. (1936). Sulla teoria di Volterra della lotta per l'esisttenza. *Giornale della Instituto Italiane Attuarti.*, **7**, 74–80.

Levins, R. (1968). *Evolution in Changing Environments*. Princeton: Princeton University Press.

Lotka, A. J. (1925). *Principles of Physical Biology*. New York: Dover.

MacArthur, R. H. & Levins, R. (1967). The limiting similarity of convergence and divergence of coexisting species. *American Naturalist*, **101**, 377–85.

May, R. M. (1972). Limit cycles in predator–prey communities. *Science*, **177**, 900–2.

Morisita, M. (1950). Dispersal and population density of a water–strider, *Gerris lacustris* L. (in Japanese). *Contributions to Physiology and Ecology*, **No. 65**.

Pimm, S. L. & Rosenzweig, M. L. (1981). Competitors and habitat use. *Oikos*, **37**, 1–6.

Pimm, S. L., Rosenzweig, M. L. & Mitchell, W. (1985). Competition and food selection: field tests of a theory. *Ecology*, **66**, 798–807.

Rosenzweig, M. L. (1971). Paradox of enrichment: destabilization of exploitation ecosystems in ecological time. *Science*, **171**, 385–7.

Rosenzweig, M. L. (1979). Optimal habitat selection in two–species competitive systems. *Fortschrift Zoologische*, **25**, 283–93.

Rosenzweig, M. L. (1981). A theory of habitat selection. *Ecology*, **62**, 327–35.

Rosenzweig, M. L. (1985). Some theoretical aspects of habitat selection. In *Habitat Selection in Birds*, ed. M. L. Cody, pp. 517–40. New York: Academic Press.

Rosenzweig, M. L. & Abramsky, Z. (1985). Detecting density-dependent habitat selection. *The American Naturalist*, **126**, 405–17.

Rosenzweig, M. L. & Abramsky, Z. (1986). Centrifugal community organization. *Oikos*, **46**, 339–48.

Rosenzweig, M. L., Abramsky, Z. & Brand, S. (1984). Estimating species interactions in heterogeneous environments. *Oikos*, **43**, 329–40.

Rosenzweig, M. L., Abramsky, Z., Kotler, B. & Mitchell, W. (1985). Can interaction coefficients be determined from census data? *Oecologia*, **66**, 194–8.

Rosenzweig, M. L. & MacArthur, R. H. (1963). Graphical representation and stability conditions of predator-prey interactions. *American Naturalist*, **97**, 209–23.

Schoener, T. W. (1974). Competition and the form of habitat shift. *Theoretical Population Biology*, **6**, 265–307.

Schroder, G. D. & Rosenzweig, M. L. (1975). Perturbation analysis of competition and overlap in habitat utilization between *Dipodomys ordii* and *Dipodomys merriami*. *Oecologia*, **19**, 9–28.

Turelli, M. (1978). A reexamination of stability in random versus deterministic environments with comments on the stochastic theory of limiting similarity. *Theoretical Population Biology*, **13**, 244–67.

Volterra, V. (1926). Variazione e fluttuzaioni del numero d'individui in specie animali conviventi. *Mem. Accad. Naz. Lincei.*, **2**, 31–113.

## R. Michael Miller

Environmental Research Division, Argonne National Laboratory

# 14 | Mycorrhizae and succession

While the existence of fungi living on or in close association with the roots of many plants has been known for many years, ecologists have only recently begun to study this association in detail and to appreciate its importance, both for the plants and fungi themselves, and for the entire ecological community and its development. This exploration has turned out to be extremely fruitful. Indeed, Mooney (1984) has suggested that the discovery of the prevalence of mycorrhizal fungi in plant communities and the growing understanding of the role they play in the life of these communities may be one of the more significant developments in ecology in the past decade.

As a result of this research it is now clear that these microscopic, below-ground organisms play an important, and in some ways vital, role in the lives of a wide variety of plant communities, ranging from the arid grasslands of the American west to the tropical forests of the Amazon basin. It has been clear for many years, for example, that mycorrhizal fungi can contribute significantly to the nutrition of many plants, especially under nutrient-poor conditions. Now it is clear that they may play a key role in the water relations of their hosts as well, and that the cost they impose on their hosts may be a critical factor in the economy of the plant in relationship to its environment. For all these reasons, mycorrhizae and the mycorrhizal association obviously play a critical role in the development of plant communities and in the dynamics of established communities.

As appreciation of the importance of this symbiotic relationship has grown during the last decade or so, research in this area has increased dramatically, and there are now numerous reviews that summarize the state of knowledge in this area (Harley & Smith, 1983; Powell & Bagyaraj, 1984; Safir, 1987). Even a cursory reading of this literature will make it clear that progress in mycorrhiza research has been extremely rapid in recent years. My purpose here, however, is not to summarize these developments, but to describe briefly how they were achieved, and especially to consider the ways in which research on mycorrhizae has profited from attempts to restore communities previously disturbed by human activities, such as mining or row-crop agriculture.

This will not be difficult because, if there is any aspect of ecological research that has profited from restoration efforts, it is mycorrhizal research. In fact, the history of research on mycorrhizae illustrates quite clearly the development of a field of study from its earliest stages as a largely descriptive endeavor to the beginnings of a more mature, experimental phase. It also illustrates the importance of the experimental – and especially the synthetic – approach as a way of working out the exact nature of an ecological relationship, defining its significance, and isolating the critical parameters that govern its functioning. Thus, scientists have been aware of the prevalence of fungal hyphae on the roots of many plants since Hartig's work in the 1840s (Trappe & Berch, 1985). Only recently, however, have they come to appreciate in any detail the importance of differences among the various kinds of mycorrhizae, the kinds of plants with which they are associated and the kinds of environments in which they occur. Interestingly, ecologists had little appreciation of the nature or importance of the mycorrhizal association until they began working with ecosystems that had been degraded by human activities, such as mining, in an attempt to reclaim or restore them as quickly, efficiently and effectively as possible. Indeed, it has been this work, far more than even the most meticulous studies of undisturbed systems, that in recent years has led to the major insights into the nature and ecological significance of the mycorrhizal association.

In short, recent research on mycorrhizae provides what may well turn out to be a classic example of restoration ecology. It illustrates, first of all, the value of working with disturbed systems, and the ways in which disturbances may elicit and bring into the foreground phenomena that might be inconspicuous in a perfectly healthy, undisturbed system. It also demonstrates clearly the ways in which attempts to restore such systems often lead to the most stringent testing of ideas about how they come together and function. Also since much of the basic work on mycorrhizae has been carried out with relatively simple, agricultural systems, it provides an excellent example of the contribution agricultural research can make to research on the more complex, naturally occurring systems, as suggested by John Harper (Ch. 3).

Finally, recent research in this area illustrates the value of the restoration approach as a way of complementing and extending research done by other means. Most studies of disturbed ecosystems, for example, have been analytical studies of population fluctuations. Such studies are not only costly, they also fail to provide an understanding of the complexity of the ecosystem as a whole (Parkinson, 1979). I have found that this problem is greatly reduced if research is organized around an attempt actually to reassemble an entire system and to get it working again. This not only provides a basis for a close interaction between various disciplines and points of view, it also helps ensure that various aspects of the problem, which may be important but tend to be overlooked, receive the attention they deserve. A case in point is the mycorrhizal association itself, and other below-ground components of terrestrial ecosystems, which have received less than their share of attention from ecologists, partly no doubt because they are simply less conspicuous than the above-ground components, and partly because they are more difficult to work with. They are, nevertheless, critically important, as research on nutrient cycling, hydrologic relations, competition and symbiotic processes amply illustrates. In this chapter, I would like to discuss briefly the development of our understanding of mycorrhizae and their ecological significance, paying special attention to the role restoration has played in this process. I will then discuss several aspects of my own work that illustrate both the value of this approach and some of its limitations.

Restoration efforts have contributed so much to our present understanding of mycorrhizae that it would be difficult to overstate their importance. At the same time it certainly is true that in many cases restoration has provided only the occasion for acquiring new ideas, information or insights that might very well have been obtained in other ways. In my discussion, I will attempt to keep this distinction in mind and to focus for the most part on examples in which I believe the restoration effort really has played a crucial role in mycorrhizal research, either by raising a new question or by providing a way of testing a hypothesis.

### Early research

Though research on mycorrhizae has increased dramatically during the past two decades, active research in this area has actually been going on for over a century, and scientists had actually learned a great deal about mycorrhizae even before the end of the last century. Indeed, in an extensive review of the first 65 years of mycorrhizal research, Mosse (1985) points out that many recent discoveries that have been hailed as new have actually been rediscoveries, often in somewhat refined form, of what had been discovered much earlier. For example, the observation that mycorrhizae flourish in nutrient-poor soils, and that they are more prevalent on plants that have coarse roots (roots with few root hairs) or otherwise reduced root systems was actually made around the turn of the century. Also during this period scientists had identified and described the three main kinds of mycorrhizae: ectomycorrhizae, or sheathing mycorrhizae, named for the mantle of fungal hyphae surrounding the root; endomycorrhizae, also referred to as vesicular–arbuscular (VA) mycorrhizae because of the type of fungal structures they produce within the root cortex; and ectendomycorrhizae, which produce both a mantle and hyphae within the root cortex, and are typically referred to as the "ericoid" type because members of this group occur exclusively on plants of the family ericaceae (heath family). Though extensive descriptive work was carried out prior to the turn of the century, these studies were generally limited in geographic scope, so that a clear understanding of which types predominate under various climatic and pedologic conditions has not been achieved until quite recently. In fact, descriptive studies are still being carried out, since the majority of the world's vegetation has still not been screened for mycorrhizal associations beyond the family level.

Studies of this kind are obviously of great importance. But they cannot tell us all we need to know about mycorrhizae. In short, they make it clear that the mycorrhizae exist, but provide relatively little information about their function or what their physiological and ecological significance might be. To answer questions about function and ecological significance, it has been necessary to resort to a more manipulative, experimental approach. Thus, characterization of the physiological relationship between mycorrhizal fungi and their hosts has depended heavily on the use of analytical, physiological techniques. Similarly, the working out of the ecological significance of the mycorrhizal association has depended on manipulative work with intact systems, and also with degraded systems, and has profited immensely from attempts not only to observe changes in systems following perturbation, but from attempts to manipulate them, to guide their development and ultimately to restore them.

Thus, what might be called a "synthetic" approach played a key role even in some of the earliest insights into mycorrhizal functioning. As early as the 1930s, for example, Hatch (1936) and McComb (1938) showed that when bare-rooted trees were transplanted into prairie soils only those that were deliberately inoculated with ectomycorrhizal fungi survived. On the basis of these findings they proposed that mycorrhizae function by increasing the absorbing surfaces of the roots, allowing for an increase in the rate of nutrient uptake.

More recently, Schramm (1966) noticed that trees needed ectomycorrhizae in order to become established on acidic mine wastes in Pennsylvania. He found that only seedlings possessing mycorrhizal fungi were growing, and pointed out that this suggested that conditions on the site had resulted in selection for mycorrhizal fungi capable of growing at relatively high temperatures. Subsequently it was found that a high substrate temperature can indeed interfere with growth of mycorrhizae, in turn affecting seedling survival (Marx, 1975). Moreover, seedlings inoculated with isolates obtained from nursery trees did not do as well as those inoculated with fungi from the harsh site. It is now known that the mycorrhizal fungus isolated from this site, *Pisolithus tinctorius*, can establish a greater abundance of hyphae during the early stages of growth than can the typical nursery isolate, *Thelephora terrestris*, and *P. tinctorius* is now used extensively in the reforestation of exposed sites.

These experiences and observations suggest the value of a restoration, or synthetic, approach to research on ecosystem function. Basically this amounts to removing elements from the system – in these cases individual trees – and then attempting to replace them. Both processes – removal and replacement – may be informative, but it may be that it is the attempt at replacement that is most likely to reveal unsuspected factors in the relationship between the element and its environment, and that replacement therefore constitutes the best test of our understanding of those factors. Descriptive studies and attempts to establish correlations may be suggestive in this regard, but they rarely provide stringent tests of ideas about function. For example, studies of derelict mesic land in many areas have established a clear correlation between successful colonization by plants and the presence of mycorrhizae (Daft & Nicolson, 1974; Daft & Hacskaylo, 1976; Danielson, 1985; Miller, 1987; Williams & Allen, 1984). But such correlations only *suggest* a role for mycorrhizae. They do not prove there is one – and if there is, they tell us little about its nature.

The earlier studies with mycorrhizae were carried out on ectomycorrhizal fungi associated with temperate forest trees, and led to the development of techniques for growing these fungi as isolates in pure culture. This analytic procedure, complemented by efforts to resynthesize the association, led to many insights into the physiology of these fungi and their relationship with their host (Harley & Smith, 1983). The value of this combination of analysis and synthesis is also borne out by more recent research on VA mycorrhizae. Although these fungi had been observed in roots since the turn of the century, until recently little was known about what organisms were involved with this association, and most scientists regarded these root-infecting fungi as weak root parasites (Schenck, 1985). Again, it was research involving isolation and reinoculation – disassembly and reassembly – that led to a revision of these views. In this case, Barbara Mosse in the UK and Jim Gerdemann in the USA used reinoculation of pot cultures by isolated spores to show conclusively that VA mycorrhizal fungi were symbiotic. This in turn provided an invaluable technique for further research on the mycorrhizal association, both *in situ* and under controlled conditions.

## Theory and reality – a case study

Ecologists studying succession on profoundly disturbed sites have found that mycorrhizae are frequently absent during the early stages of succession, and that many of the plants characteristic of these stages are either nonmycorrhizal or facultatively mycorrhizal species (Janos, 1980; Reeves *et al.*, 1979). Since reinoculation experiments have repeatedly shown that the uptake of mineral ions by plants increases and water relations are improved following inoculation, it seems likely that this is why mycorrhizae enable plants to establish themselves on these harsh sites. On the basis of this and other evidence, Reeves *et al.* suggested that if mycorrhizae do play a critical role in succession, and if succession beyond the early seral stages actually depends on the presence of mycorrhizae, which in turn facilitate establishment of mycorrhizal plants, then it might be possible to skip stages in succession by adding mycorrhizal fungi. This, of course, would be of great value in attempts to restore native, late-seral vegetation on sites as quickly as possible. It would also be of considerable theoretical interest as a test of ideas about the mechanics of succession based exclusively on descriptive and analytical studies.

In fact, our experiments to test this idea have produced negative results. When we initiated studies on the effects of topsoil handling procedures on mycorrhizal fungi propagule survival and plant establishment at a surface mine site in Wyoming, we found that non-mycorrhizal plants persisted on the site, even though mycorrhizal propagule levels in the soil were similar to those in an adjacent native community (Miller, 1979). Although this study lacked an appropriate control (plots completely lacking mycorrhizal fungi being nearly impossible to maintain in the field), the results strongly suggested that mycorrhizal fungi were in fact not limiting succession and that the accepted idea that mycorrhizae directly influence succession in a simple way is naive. In fact their role is quite complex, as will be shown below.

I began work on the restoration of arid lands disturbed by coal mining in the American west in 1975. At that time it was generally believed that, because of their low diversity and unpredictable precipitation and temperature regimes, these communities are quite fragile, and there was considerable skepticism about the possibility of restoring them. Although mycorrhizae were known to be present in plants occurring in this region, little ecologically oriented research had been done on them, and the ways in which they were affected by soil disturbance were a complete mystery. In mining, the majority of salvaged soils were stockpiled until needed. This was necessary because extraction typically occurs at a faster rate than reclamation. One major concern had to do with the way soil and soil organisms change during stockpiling. Until our study, only guesses as to what was occurring were available, but it was clear that stored topsoil often did not perform as well as freshly removed topsoil when reapplied on overburden. Also, the usual approach to handling topsoil at that time was to treat it as a mineral, concentrating on its physical and chemical properties and ignoring its biological attributes.

In an attempt to learn something about the resilience of profoundly disturbed soils, and especially about how the biological properties of soil are affected by different handling procedures, we began with a simple experiment, comparing the development of vegetation on soil that had been removed from a native community and immediately spread on a disturbed site with development on soil from a similar community that had been stockpiled (in heaps up to 15 m deep) for two years before respreading (Miller, 1984).

Though initially designed merely to evaluate revegetation success, this simple experiment led to numerous insights into the factors influencing the recovery of vegetation on disturbed sites. Our first observation was essentially what we had expected: both soils were rapidly invaded by exotic annual species, mostly annual members of the Chenopodiaceae, notably *Halogeton glomeratus* and *Salsola kali*. In addition, a number of natives appeared on both sites, arising from seed in the soil. As might have been expected, however, there were far fewer natives on the soil that had previously been stockpiled than on the freshly spread, direct-applied topsoil, suggesting that seed viability declined during storage. (An alternative explanation is that there was something in the soil of the stored-applied treatment that interfered with germination of these seeds (Jastrow *et al.*, 1984).)

Studying alterations in the mycorrhizal components of these communities, however, turned out to be far more complex. As these experiments got under way, we learned that little was known about the presence of mycorrhizae in the native communities we were trying to restore, and essentially nothing at all was known about how mycorrhizal fungi respond to disturbance. Our experiments, however, were quite informative. For one thing, we found that, although most members of the Chenopodiaceae are non-mycorrhizal, specimens of certain species in this family often possessed mycorrhizae when they appeared on our plots (Miller, 1979; Miller, Moorman & Schmidt, 1983). Also, even though repeatedly seeded onto these sites, and known to flourish on similar sites prior to disturbance, the mycorrhizal wheatgrasses (*Agropyron smithii* and *A. dasystachyum*) failed to establish. Moreover, in the absence of large populations of mycorrhizal plants, the numbers of mycorrhizal propagules in the soil were decreasing steadily (Fig. 14.1). Since many of the species of the native community require mycorrhizae, we began to suspect that the moving and stockpiling of soil had set changes in the mycorrhizae in the soil in motion that were not being reversed simply as a result of respreading the soil and adding grass seed. At this point we thought that the system was deteriorating ecologically, and we predicted that the result would be a community with fewer species than the native community, and that many of the species present would be exotics.

Then, quite unexpectedly, five years after the spreading of the soil and the first seeding of grasses, the mycorrhizal infection potential of the soil increased for both the direct-applied and stored-applied soil application treatments. Surprisingly, the infection potentials for the direct-applied soils were actually higher than those found in soils of adjacent native communities, whereas the infection potentials of stored-applied soils were still significantly lower than those of soils occurring in the native communities. In an attempt to discover what was going on here we relied on past measurements of nutrients and changes in these nutrients over

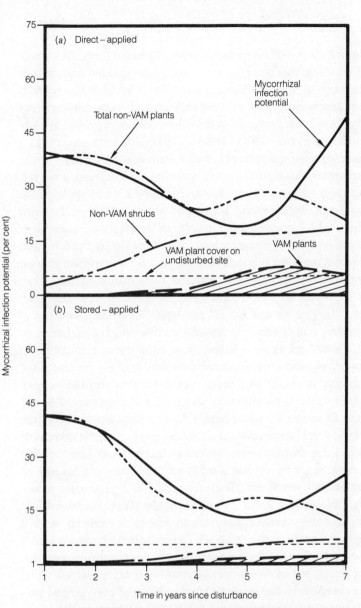

Fig. 14.1. Mycorrhizal infection potentials at first declined steadily on disturbed sites where soil was spread immediately following removal from a mined area, then increased dramatically after the fifth year, when the density of mycorrhizal (VAM) plants emerging from viable seeds remaining in the soil, as well as from applied seed, approached that typical of natural communities. Since many late successional plants of this cold desert community are obligate mycotrophs, which require mycorrhizae to grow under these conditions, this represents an important step toward reestablishment of the native community on this site.

On sites covered with soil that had been stockpiled for two years, in contrast, fewer mycorrhizal plants appear, mycorrhizal infection levels remain low, and, though the sequence of events appears to be similar, populations remain at a much reduced level.

time. Similarly, we also relied on measurements of precipitation. As a result of these efforts we found that changes in the carbon-to-nitrogen and carbon-to-phosphorus ratios in the soil following disturbance led after five years to conditions that (apparently) favored the replacement of nonmycorrhizal ruderals such as *Halogeton* and *Salsola* by *Atriplex* shrubs, mainly *A. confertifolia* (Jastrow *et al.*, 1984; Miller, 1984). Moreover, there was a close correlation between the presence of this plant and an increase in the number of mycorrhizal propagules in these soils. We soon realized, however, that this association was not due to the presence of the shrub itself, but to a second association, between the shrub and the volunteer and seeded grasses. In fact, while grasses fared poorly in most areas, they were establishing quite well in the immediate vicinity of the *Atriplex* shrubs, and it was these grasses, and not the shrubs, that were actually responsible for the increase in mycorrhizal propagule levels in these disturbed soils.

This led us to search for factors that might account for the close association between the grasses and the *Atriplex* spp. There are a number of obvious possibilities, one of which is that the shrubs, which stand more or less intact all winter, act as snowfences, increasing the accumulation of snow near the shrub and creating microsites that actually receive more water than adjacent areas. In fact, there is evidence that shrub cover is a significant factor influencing snow capture in semi-arid ecosystems (West & Caldwell, 1983). Of course it is also possible that the snowfence effect of the shrub might encourage accumulation of seeds, spores, etc. To determine whether any of these factors were significant in our case, two growing seasons after disturbance we erected a set of artificial snow catchments on the stored-applied soil treatment. These consisted of fine-gauged screen fences 1 m high, placed in a 25 m × 10 m rectangle. (This experiment is in itself an example of the synthetic approach to ecological research, since it involved setting up "artificial shrub islands" in order to determine whether the factors responsible for the effects we were seeing around shrubs were due to biological or merely to physical factors.) In fact, we obtained evidence that physical factors were involved because plant cover and mycorrhizal propagule levels were significantly greater inside the catchments than they were outside them. Both plant cover and mycorrhizal propagule levels were greater in the catchments than in nearby undisturbed communities.

Finally, as a result of observing and tinkering with a restoration project set up to test one well-defined hypothesis and carrying out the appropriate experiments as questions arose, we have been able to piece together a reasonably clear idea of what happens during succession in this particular arid, shrub community. We have also been able to identify at least some of the key factors influencing this process, and are therefore now in a better position to propose ways of guiding it.

First, it appears the site is invaded by opportunistic, ruderal exotics. None of these requires mycorrhizae, and although several species may become infected by mycorrhizae if they are growing near mycorrhizal plants, for the most part species appearing early on these sites remain non-mycorrhizal. As a result, there is a progressive decline in the number of viable mycorrhizae propagules existing in the soil (Fig. 14.1). This strongly suggests that the site may be on its way to developing a community of low diversity, made up largely of non-mycorrhizal species and quite unlike the native, model community. We find, however, that this can be altered quite dramatically by a simple sequence of events triggered by the presence of shrubs emerging from seeds in the soil bank which, almost certainly acting as catchments for snow, and probably also for seeds and mycorrhizal spores, favor the development of mycorrhizal species, principally grasses. These then act to increase the population of mycorrhizae in the soil, in this way facilitating change in the direction of a diverse community rich in native species. In contrast, on stockpiled topsoil this process is less likely to occur, not because of the loss of mycorrhizal propagules during stockpiling (Miller, Carnes & Moorman, 1985), but because of depletion of the seedbank. As a result the sequence of changes on these sites differed dramatically from those that occurred on sites with direct-applied soil.

While this picture emerged as a result of a variety of observations and techniques employed as we monitored the recovery of this system and attempted to account for its progress, it is important to note that one of the most significant findings – the discovery of the subtle dependency between shrubs and grasses – was achieved by investigating different methods used in handling topsoils during surface mining. The primary difference between these handling procedures was simply that in one case the soil was applied directly, and in the other it was stockpiled before being applied over regraded overburden. Though not measured directly, a major result of stockpiling topsoil was an apparent reduction in the viability of seeds in the soil. This factor, together with the reduction in mycorrhizal propagule viability associated with storage, and amplified by the senescence of propagules in the absence of mycorrhizal host plants, has led to a valuable insight into factors influencing successional trends in this community. The most obvious difference between the two treatments was related to shrub establishment. Few shrubs of *A. confertifolia* became established on the stockpiled soils, whereas these shrubs are a major component of the community that has developed on the direct-applied soils. This difference in seedbank viability, along with the abnormally low annual precipitation that occurred during the study, probably increased the importance of the shrubs for both grass establishment and mycorrhizal propagule survival. In any event, while mycorrhizal plants have developed at the expense of nonmycorrhizal plants on sites with direct-applied soil, obligate mycotrophs have not colonized these sites. Since these obligately mycorrhizal plants comprise a major portion of the native plant community, this means that the biodiversity of these sites is still significantly lower than that of nearby native communities.

The sequence of events described above could not have been discovered through dissection of the undisturbed community. In fact, the vegetation changes we saw on our restoration area differed dramatically from those in nearby natural communities, where there was actually a negative association between grass cover and mycorrhizal propagule levels. The most likely explanation for this appears to be related to the growth habit of these plants. The grasses at this site are rhizomatous and so able to grow away from the "nurse" shrub in search of nutrients. This tendency might well mask an historical association between grass and *Atriplex* if only later successional communities were studied.

### Further questions

At this point in our work it is easy to point out issues in mycorrhizal ecology that might be studied using the techniques of restoration ecology. Here are a few examples.

#### The ecological importance of hyphal interconnections

There is considerable evidence from analytical studies (Read, Francis & Findlay, 1985) that hyphal connections between plants persist following infection by mycorrhizae, and that nutrients are exchanged through them. What remains unclear, however, is the extent to which this is of ecological significance. For example, hyphal linkages might permit the survival of shaded seedlings that might otherwise perish, and this might obviously have an important influence on the development and final composition of a plant community. Probably the only means of determining the role of these connections is an approach based on recreating increasingly complex associations, including various plant-to-plant, plant-to-fungus and fungus-to-fungus combinations.

## Role of mycorrhizae in formation of soil aggregrates

Our experiences with the tallgrass prairie being restored in the accelerator ring at Fermi National Laboratory near Chicago have led to the recognition that mycorrhizae on the roots of prairie plants bring about recovery of the crumb structure of degraded agricultural soils more rapidly than any other plant community currently used in intercropping. This observation, which depends entirely on the attempts to restore prairie on degraded soils, has important implications. For one thing it suggests that prairie restoration may be a valuable, and even an indispensable technique for rejuvenating degraded agricultural soils. On the more purely ecological side it provides valuable information about the formation of prairie soils.

## The physiological cost of mycorrhizae and its ecological implications

How much energy does it cost a host plant to acquire nutrients *via* mycorrhizae? The cost to individual plants can be determined analytically by physiological studies. But the ecological consequences of this on the community level have to be determined by other means, the most obvious of these being the deliberate attempt to recreate communities with and without the mycorrhizal association. These kinds of studies would require the fumigation of soil to eliminate the mycorrhizal fungus. However, fumigation has its drawbacks. An alternative approach would be to use soil not previously occupied by mycorrhizal plants, since these soils would be void of any significant mycorrhizal propagule reserve.

### Acknowledgments

Support for both the research described and the preparation of this chapter was provided by the US Department of Energy's Office of Health and Environmental Research, under contract No. W–31–109–ENG–38 to Argonne National Laboratory.

### References

Daft, M. J. & Hacskaylo, E. (1976). Arbuscular mycorrhizas in anthracite and bituminous coal wastes of Pennsylvania. *Journal of Applied Ecology*, **13**, 523–31.
Daft, M. J. & Nicolson, T. H. (1974). Arbuscular mycorrhiza in plants colonizing coal wastes in Scotland. *New Phytology*. **73**, 1129–38.
Danielson, R. M. (1985). Mycorrhizae and reclamation of stressed terrestrial

environments. In *Soil Reclamation Processes*, ed. R. L. Tate & D. A. Klein, pp. 173–201. New York: Marcel Dekker.

Harley, J. L. & Smith, S. E. (1983). *Mycorrhizal Symbiosis*. London: Academic Press.

Hatch, A. B. (1936). The role of mycorrhizae in afforestation. *Journal of Forestry*, **34**, 22–9.

Janos, D. P. (1980). Mycorrhizae influence tropical succession. *Biotropica* (Tropical Succession), **12**, 56–64.

Jastrow, J. D., Miller, R. M., Rabatin, R. C. & Hinchman, R. H. (1984). Revegetation of disturbed land in arid ecosystems. In *Ecological Studies of Disturbed Landscapes*, Ch. 2. DOE/NBM–5009372.

McComb, A. L. (1938). The relationship between mycorrhizae and the development and nutrient absorption of pine seedlings in a prairie nursery. *Journal of Forestry*, **36**, 1148–54.

Marx, D. H. (1975). Mycorrhizae and the establishment of trees on strip-mined land. *Ohio Journal of Sciences*, **75**, 288–97.

Miller, R. M. (1987). The ecology of vesicular–arbuscular mycorrhizae in grass- and shrublands. In *The Ecophysiology of Vesicular–Arbuscular Mycorrhizal Plants*, ed. G. R. Safir. Boca Raton: CRC Press.

Miller, R. M. (1984). Microbial Ecology and Nutrient Cycling in Disturbed Arid Ecosystems. In *Ecological Studies of Disturbed Landscapes*, Ch. 3. DOE/NBM–500-9372.

Miller, R. M. (1987). The ecology of vesicular–arbuscular mycorrhizae in grass- and shrublands. In *The Ecophysiology of Vesicular–Arbuscular Mycorrhizal Plants*, ed. G. R. Safir. Boca Raton: CRC Press.

Miller, R. M., Carnes, B. A. & Moorman, T. B. (1985). Factors influencing survival of vesicular–arbuscular mycorrhiza propagules during topsoil storage. *Journal of Applied Ecology*, **22**, 259–66.

Miller, R. M., Moorman, T. B. & Schmidt, S. K. (1983). Interspecific plant association effects on vesicular–arbuscular mycorrhiza occurrence in *Atriplex confertifolia*. *New Phytology*, **95**, 241–6.

Mooney, H. A. (1984). Progress and promise in plant physiological ecology. In *Trends in Ecological Research for the 1980s*, pp. 5–17. New York: Plenum Press.

Mosse, B. (1985). Endomycorrhiza (1885–1950): The Dawn and the Middle Ages. In *Proceedings of the 6th North American Conference on Mycorrhizae*, pp. 48–55. Oregon: Forest Research Laboratories, Oregon State University.

Parkinson, D. (1979). Microbes, mycorrhizae and mine spoil. In *Ecology and Coal Resource Development*, Vol. 2, ed. M. K. Wali, pp. 634–42. New York: Pergamon Press.

Powell. C. L. & Bagyaraj, D. J. (1984). *VA Mycorrhiza*. Boca Raton: CRC Press.

Read, D. J., Francis, R. & Findlay, R. D. (1985). Mycorrhizal mycelia and nutrient cycling in plant communities. In *Ecological Interactions in Soil*, ed. A. H. Fitter, pp. 193–217. Oxford: Blackwell Scientific.

Reeves, F. B., Wagner, D., Moorman, T. & Kiel, J. (1979). The role of endomycorrhizae in revegetation practices in the semi-arid. I. A comparison of incidence of mycorrhizae in severely disturbed vs natural environments. *American Journal of Botany*, **66**, 6–13.

Safir, G. R. (1986). *The Ecophysiology of VA Mycorrhizal Plants*. Boca Raton: CRC Press.

Schenck, N. C. (1985). Vesicular–arbuscular mycorrhizal fungi: 1950 to the present – the era of enlightenment. In *Proceedings of the 6th North American Conference on Mycorrhizae*, pp. 56–60. Oregon: Forest Research Laboratories, Oregon State University.

Schramm, J. E. (1966). Plant colonization studies on black wastes from anthracite mining in Pennsylvania. *Transcriptions of the American Philosophy Society*, **56**, 1–194.

Trappe, J. M. & Berch, S. M. (1985). The prehistory of mycorrhizae: A. B. Frank's predecessors. In *Proceedings of the the 6th North American Conference on Mycorrhizae*, pp. 2–11. Oregon: Forest Research Laboratories, Oregon State University.

West, N. E. & Caldwell, M. M. (1983). Snow as a factor in salt desert shrub vegetation patterns in Curlew Valley, Utah. *The American Midland Naturalist*, **109**, 376–9.

Williams, S. E. & Allen, M. F. (1984). VA Mycorrhizae and Reclamation of Arid and Semi-Arid Lands. In *Proceedings of the Conference*, August 17–19, 1982. University of Wyoming Agriculture Experiment Station Report No. SA1261.

*James A. MacMahon*

Department of Biology, Utah State University, Logan, Utah,

# 15 Disturbed lands and ecological theory: an essay about a mutualistic association

### General introduction

Once a portion of the landscape is disturbed, its regeneration toward a "natural" state involves phenomena and processes that are the bases of a significant portion of ecological theory, or, as some might have it, ecological speculation.

Management of disturbed lands to hasten their regeneration is, in its simplest interpretation, merely the management of ecological processes for a specific purpose. Sometimes management schemes are developed to shorten the time period to complete the regeneration effort, achieving as an endpoint an ecosystem that closely resembles surrounding, undisturbed reference areas. Under certain circumstances an endpoint is chosen that does not represent the natural, adjacent landscapes. In either case, knowledge of the ecological theory that pertains to community regeneration may permit the development of a more effective management program – that is, one that is less expensive, can be implemented faster, and gives a more desirable final result. Obviously, if attention to ecological theory can achieve all of these laudable goals, it deserves our scrutiny.

Because lands disturbed by human activities are generally subject to one or another set of regulations, they are often the focus of considerable research. Such research would not be possible were it not for the implied economic importance of compliance to regulations that are developed to mitigate the effects of the disturbances. Moreover, the results of this research and of the management applied to such sites can provide a data base to assess the robustness, and even correctness, of ecological theory. Indeed, disturbance of ecosystems by humans often offers such a stark contrast to natural disturbances, because of its intensity, that we may more clearly see and understand ecological processes and their linkages by studying such areas than by studying natural ecosystems. In addition the short time period involved in reclamation permits us to observe results that would span tens to hundreds of years of natural regeneration.

Since reclamation is really a form of "applied" succession, this personal narrative will address several broad ecological constructs related to the ubiquitous process of succession. It will emphasize both the value of ecological theory to management programs, and a consideration of the impact that management studies can have on ecological theory. Since my experiences with management have been confined to reclamation of land that was surface-mined for coal, the catholicity of my comments is limited. The practical value of such work is obvious, however, since 930,000 ha of the USA have been disturbed by surface-coal mining, and since the conterminous 48 states still contain four million ha of strippable reserves, spread over 28 states (Dvorak, 1984). Of even broader value, however, are certain lessons we have learned from this work, many of which may have even wider applicability.

The examples for certain statements will be drawn from our own work on a strip mine in Wyoming or on the work we have been doing on Mount St Helens, a site we believe has many similarities to strip mines. This provincial approach is not meant to demean a rich extant literature but simply to provide an account of the specific studies that have been the basis of my personal experiences, and thus my opinions.

### Introduction to succession

Clements (1916) codified what was known about the general phenomenon of succession. His view was that succession represented a "universal law" that accounts for the predictable, directional change in the nature of a disturbed place that is caused by the organisms at that place and that eventually leads to a stable community, termed the climax. If a particular piece of land was composed of raw geological materials, Clements termed the process primary succession and implied it would take a very long time to reach stabilization. In contrast, if the piece of land had previously supported a community, and had been disturbed, then the same climax endpoint would be approached, but at a much faster rate, and the process is referred to as secondary succession.

Today, both the absolute directionality of the process, and the certainty of one, invariant climax state are viewed as over simplifications. However, most ecologists, whether they realize it or not, still adhere to a Clementsian notion that beginning with bare rock or barren soil, given enough time, a plant community and some animal associates will occupy any site. In addition, it is still widely assumed that the mixtures of species will be virtually the same as, or at least highly reminiscent of, mixtures found in the general area of the site.

To account for the changes observed during succession, Clements categorized the processes responsible for succession and discussed variants of these processes that might occur under certain specific conditions of disturbance or environment. A short summary of his views and the specific processes he recognized suggests that, in general, the process of succession is initiated after a disturbance, the nudation process. Some organisms – termed "residuals" – might survive that event. These, in time, are joined by nearby or highly mobile species that can easily reach the disturbed site.

Both the migrants and the residuals must establish populations that can flourish (the process of ecesis) otherwise they would die off or emigrate and would have little influence on that site. As species become established they interact with other species. Clements emphasized competition as the main form of such interaction; however, we know now that any species–species interaction (for example, mutualisms, predator–prey relations, etc.) can be important to the developing community. It is also clear that the very presence of species alters the environment – plants, for example, cast shade, changing soil temperatures and may simultaneously add organic matter to the soil. Clements termed this phenomenon "reaction" and thought it created a new environment, one that was no longer suitable for the very species that were causing the changes. Under such circumstances a different set of migrants, or even residuals, would be favored, and the composition of the community would change. The cycle would repeat itself until a group of species was established, and an environment was generated, that were in perfect equilibrium. This state of stabilization, usually termed the climax, perpetuates itself until a new denuding event occurs. Following such an event the same processes of migration, ecesis, reaction and the development of biotic interactions would operate through time to bring the site back to the inevitable climax, a condition regarded as being independent of the nature of the denuding force.

To date many variants of Clements's theme have been proposed, though no one general model of succession has enjoyed complete acceptance among ecologists. Despite the plethora of extant opinions concerning the details of succession, Clements's simple scheme of processes has heuristic value and I will use these terms to discuss the components of succession in the context of managed ecosystem regeneration.

### Nudation – the process of disturbance

Whatever their effect on the end result of a succession, the details of the disturbance process do set the stage for all subsequent events in either managed or natural revegetation and refaunation. The type of disturbance, its extent, season of occurrence and intensity, as well as the type of community affected and numerous other factors clearly do influence the post-disturbance trajectory of recovery.

A few examples will clearly elucidate these relationships. A crown fire in a Minnesota conifer forest, on a cool spring day, is different in its effects from a ground fire on that day, or from a crown fire during a different season or on a different type of day in the same season. The type and intensity of fire can determine which species of plants and animals survive and act as the residuals or founders for the early successional stages. Specific reasons for these effects include the fact that the serotinous (closed) cones of a variety of conifers require fires of specific intensities before they will open. Seeds in the soil reserves may or may not survive depending on the nature of the fire. The same is true for animals. Even the architectural structure of the post-fire community can influence reestablishment of species. A forest that is leveled does not attract many birds. In contrast, the standing dead boles of trees provide sites for numerous bird species that are associated with "snags". Interestingly, standing snags can also increase the probability of the occurrence of a later fire because they represent a form of fuel build-up. Thus, the nature of one disturbance can actually influence the probability of the occurrence of another disturbance at a later time!

In addition to these "biological" influences, the denuding event also sets the abiotic stage. For example, the type of fire determines such characteristics as the carbon:nitrogen ratio of the soil and the permeability of the soil surface to water and nutrients.

It should be clear that it is the residuals, both biotic and abiotic, that are the legacy of the nudation process, and it is these residuals that influence ecosystem regeneration. This influence has often been ignored in ecology, or at least given scant attention. But studies of mined sites, even more than studies of native ecosystems, have drawn ecologist's attention to the importance of the nudation process and the consequent residuals of that process.

In the American West, for example, abandoned sites of all sorts, including strip mines, are colonized by a host of "weeds". These are often non-native annuals such as Russian thistle (*Salsola kali*) that can retard the growth of species planted to reclaim the land, and as such are economically significant "pests." The source of these plants is frequently the soil seed reserves (residuals), though some species, including Russian thistle, have highly effective dispersal mechanisms, making them common early migrants.

This negative effect of residuals contrasts with the positive effect of other desirable residual species that are influenced by specific mining practices. Topsoil that has been stripped prior to mining is often stored in piles, later to be replaced on the reclaimed site. It is now well known that native seeds and the spores of beneficial fungi deteriorate in storage, especially under moist soil conditions. To offset this, some mines strip the soil from one site and immediately place it on another site that is being reclaimed nearby. This procedure often results in the rapid development of an effective cover of plants developing from residual seeds, and thus reduces the cost of providing a non-native cover crop, that will later be destroyed.

This experience clearly demonstrates that nudation and the subsequent residuals *can* alter the course of succession. Industry now successfully manages sites to enhance survivorship of residuals that aid the revegetation process. Studies of such sites allow ecologists to examine the exact mechanisms underlying the influences of the various residuals, and these suggest parallels to be explored in native systems.

## Migration – The dispersal of the new colonizers

The movement or transport of organisms, their dispersal, is important to the study of the reclamation process. Migrants, along with residuals, form the foundation of any developing community.

The probability of a species being a migrant onto a particular site depends on a number of factors, but especially on the distance of the species from the site and the mobility of that species. The obvious relationship between proximity and the success of colonization need not be detailed. However, note that proximity is measured in relative, not absolute, distances. What is "close" for one species could be a great distance for another. This is in part due to the fact that there are two general types of migration, active and passive. In active migration the ability and/or tendency of a species to move plays a significant role determining that species' probability of migration.

There is an important caveat in this statement. Some species have the capacity to move great distances, but may limit their movements because of barriers that are trivial in terms of distances involved, but absolute in a behavioral or physiological sense. Some tropical forest-dwelling birds such as antbirds and toucans, for example, can fly great distances, but seem not to cross open water areas of a few tens of meters (Brown & Gibson, 1983).

In the case of passive migration many other factors, biotic and abiotic, may be at work. For example, some passive dispersers, such as many plants with fleshy fruits, are dispersed by animal consumers. Their migration distances, therefore, depend on who eats them. A bird and a mouse eating the same fruit might move the indigestible seed quite different distances. The seeds of a wind-dispersed plant may move different distances depending on the weather conditions on the day the seeds are ready for transport. In experiments with cheatgrass (*Bromus tectorum*) on our mining site, it is clear that spring movements are over shorter distances than those of the summer, when strong, turbulent winds can lift seeds into the air stream rather than just pushing them along the ground.

Obviously, getting organisms to a site is a process that has both chance and predictable deterministic elements. In a managed situation the "planter" is attempting to remove the guesswork because that person is an agent of passive dispersal, regardless of the dispersal type of the species.

Leaving mature, seed-producing trees in an area that is otherwise a clearcut obviously reduces the dispersal distance for seeds. Adjusting the size and shape of an area to be disturbed can also aid recolonization. Long narrow areas have a large edge area ratio, and are therefore likely to be recolonized more quickly than wider areas. Using knowledge of the direction and force of prevailing winds when choosing sites to leave as seed source islands can also save both time and money. For species that are passively moved by animals, a manager needs to find ways to encourage the animals to return to the recolonizing site. The plants will follow with them. Envision a mined site with no vertical architectural component – that is, no trees or tree-like structures. Such a site will not provide roosting sites where birds might rest and defecate seeds of desirable colonists. On such a site, a few artificial perches might aid reestablishment of key species. An added bonus is that some seeds are difficult to germinate unless they have passed through a consumer's gut tract, so drawing the consumer to a site elegantly solves both the migration and germination problems.

All these observations obviously have important practical implications. Equally important, however, is the fact that our studies of mined sites have helped us to better understand the whole process of migration and its role in community reassembly. The large expanses of open mined sites are natural laboratories that enable us to look at the relationship between distance from a source of propagules and the structure of a developing community. In many sites there are no residuals because mine spoils are sterile. Thus, plant and animal establishment is related to dispersal, not the establishment of residuals. In this stark scenario the importance of particular life history strategies of plants and animals can be studied with reference to their influence on ecosystem organization. This is an important area of ecological theory that has been the subject of much speculation, but few data have been forthcoming. Studies of mined sites are allowing us to evaluate postulates such as those that are so abundant in Grime's (1979) seminal book on vegetation processes.

### Ecesis – the establishment of the biota

Just because a species survives a particular denuding event or is able to migrate to a particular spot, there is no assurance that it will be able to survive on that site. Ecesis, the establishment of the individuals or propagules of that species, is also required. Mount St. Helens provides a clear case in point. Pocket gophers (*Thomomys*) and moles (*Scapanus*) survived the critical volcanic blast in some places because both species were protected by their fossorial (belowground) habits. In some cases, however, the moles later died off and failed to establish because the invertebrates they require as food did not survive. In contrast, pocket gophers subsisted on residual bulbs, roots, and rhizomes of plants. Thus, the initial survivorship and subsequent persistence at the same site were quite different, even for mammals that are superficially similar (Anderson & MacMahon, 1985a).

The importance of ecesis as a process separate from other aspects of succession is well illustrated on mined sites. Seeds are often spread on or drilled into soils that do not permit establishment. Perhaps soil water potential is too low for germination and growth, or maybe nitrogen levels are too high or too low. For an animal the food supply may be sufficient, but some key requirement in the life cycle is missing, perhaps a nest site or a suitable microhabitat in which to spend a winter of torpor.

In the west, recently reclaimed mined sites frequently harbor mammal populations that are more dense than those in the surrounding native vegetation. Such sites make good feeding areas for golden eagles (*Aquila chrysaetos*), but the eagles often do not use these areas because of the absence of suitable roosting and nesting sites such as cliffs. When recontouring the land following strip mining, some mine operators have found that leaving some of the mines's high wall, a vertical section of the mined pit, in place will act as a rock outcrop suitable for eagle nesting. Eagles or hawks that could establish in the presence of a high wall might act as a check on granivorous rodents that attain high population densities on seeded, early reclamation sites, and, much to the chagrin of the manager, feed on the seeds on which reclamation depends (Parmenter & MacMahon, 1983).

Since disturbed mined sites are often resoiled with material that has been homogenized during the processes of stripping, hauling, storage and reapplication, and since these are often spread over topographic – and therefore microenvironmental – gradients, ecologists can use mined sites as a gigantic "greenhouse" experiment. For plants, for example, the soil substrate may be uniform, but the depth and persistence of the snowpack may vary appreciably from point to point. Multivariate statistical analyses of the patterns of plants across mined sites can help us to elucidate exactly which environmental factors are most important for the growth and development of particular plants or animals (compare Chs 12 and 17). These data aid ecologists as they attempt to dissect out the causal mechanisms of community development. These same data allow us to better manage desirable or undesirable species in nature. Knowing the requirements for a species' establishment increases our chances of controlling that species in nature. The mining scenario allows us to address the problem of establishment at the level of a landscape rather than relying on experiments in pots in a greenhouse. It is becoming increasingly clear that in ecology there is a level of integration of biological systems – the landscape level – that has characteristics best studied at that scale and not at one of finer resolution. This is the whole basis of the emerging discipline of landscape ecology (Naveh & Lieberman, 1983). Disturbed sites bring principles of landscape ecology into focus and provide excellent laboratories for their development and appraisal.

### Biotic interactions – the biological plexus

Species interact with one another with consequences that can be either positive, neutral, or negative for each of the interacting species. Competition, the species–species interaction that Clements thought was so important for species during succession, is a negative interaction because both species are competing for a limiting resource. The interaction between a plant and its pollinator is just as important in structuring an ecosystem, but represents a benefit for both interacting species.

When an interaction is obligatory for one or both interactants, the establishment of the relationship is a requisite for ecesis.

In the last section, discussing the development of habitat for eagle population establishment, I wrote about eagles eating mammals. Obviously, that example contains, in addition to an ecesis component, a biotic interaction portion. The eagle and the rodent are predator and prey, and the presence of the eagle may prevent establishment of the prey. This underlines the somewhat artificial boundaries of the processes I have chosen to discuss, but it also illustrates the way in which a single species may interact with a variety of other species in ways that are difficult to characterize in a tidy manner.

In the severely disturbed areas of Mount St Helens, plants that have dispersed there encounter pumaceous soils that are low in nutrients and water-holding capacity. Often migrants that germinate in the St Helens soils cannot persist to form self-reproducing populations. In many areas this may be due, in part, to a missing biotic interaction. Over 90 per cent of the world's vascular plant species form an association with fungi that is known as a mycorrhiza. The association allows the fungus to acquire food from the plant. The myceliar structure of the fungus may help the plant acquire nutrients, especially phosphorus and water, when they are in scant supply. The spores of the fungus are carried by wind and animal vectors, but have not infected plants in certain large (hundreds of km$^2$) areas near the volcano (Allen & MacMahon, 1984). In other areas where the volcanic materials form a thinner veneer over buried soils, the soil-mixing action of mammals has brought residual spores into association with plant roots. These plants are mycorrhizal, and they seem to have a better chance of establishment than nearby individuals of the same species that have not formed the association (Anderson & MacMahon, 1985b).

This same biotic association is important on mined lands. One of the advantages of the direct placement of topsoils, as opposed to soil storage, is that the directly placed soils have more viable spores than those soils that were stored. Consequently plants might do better in these biologically more active soils. The development of the mycorrhizal symbiosis is an important phenomenon in nature, and the studies we have conducted on mined lands have put this into perspective in our thinking. Indeed, attempting to explain the patterns of apparent plant succession on mine spoils has been the key to elucidating the overall importance of mycorrhizae in all of our ecological studies. Because of the mined land studies, we have paid particular attention to how the fungus actually migrates or survives as a residual and how a spore can ultimately form the mycorrhiza with the plant. These processes have implications for the establishment of plants on any site anywhere in the world (for further discussion see Ch. 14).

### Reaction – the organisms change their environment

Plants grow and shed their leaves and roots, resulting in increases in soil organic matter. Microbes fix nitrogen, altering soil chemistry. The burrowing activities of animals change soil structure, which in turn alters soil water-holding capacity and aeration. The presence of a plant alters the soil surface temperature. This temperature change may, in significant measure, alter the composition of the soil animal community. All of these, and many other biotically induced changes, are part of the reaction process. In a literal sense the organisms are changing the environment in which they live, perhaps even to the point of creating one that is unsuitable for their own progeny. Plants requiring high light intensities often cannot establish in the shade of their parents. Any of these environmental changes, as noted earlier, may alter the composition of the community. A different set of migrants might establish under the new regime. Perhaps some dormant residual could now establish after being repressed by earlier conditions. At any rate, once a new suite of species establishes they may create an environment that they themselves cannot endure, and yet another species constellation will replace them. This repeated sequence of induced environmental changes and subsequent species turnover has been the basis of the more classical view of an orderly succession, referred to by terms such as relay floristics or the facilitation model (Connell & Slatyer, 1977). Eventually, the result is a persistent mix of species, one that generally perpetuates itself over a long period of time. The attainment of this "stable" climax admixture is called stabilization and is the endpoint sought in most restoration or reclamation schemes.

Much of what we do as reclamationists is, in a real sense, directed to the regulation of the reaction process. Planting of cover crops, fertilization and even irrigation are activities meant to alter soil properties (reaction) so that late succession plants can undergo ecesis.

One important aspect of reaction that is often overlooked is the development of the general form of the community, irrespective of the species composition. This includes both the architecture of the above- and below-ground portions of plants (the vertical aspect) and the horizontal distribution (dispersion) of plants, animals and soil constituents (see also Ch. 22).

Examples of the importance of vertical and horizontal architecture will make this point clear (Fig. 15.1). The optimum dispersion pattern of plants in arid-land communities has been the topic of some debate. Many authors felt that plants should be regularly spaced – like trees in an orchard – so that individuals could avoid competition for water and nutrients by having their own exclusive area to live in. Actual measures in the field showed, however, that while some plants were regularly dispersed, many species were clumped. These clumps contained not one, but a variety of species.

To explain this clumping phenomenon it was hypothesized that the

*Fig. 15.1.* Bleak aspect of reclamation sites where the work described in the accompanying chapter was carried out is clear from this photo of a planting crew in action. The crew hand-planted a total of 120,000 containerized shrub tublings as part of experiments to determine the influence of dispersion patterns on the speed of recovery. The site shown is the Kemmerer Coal Mine in Wyoming.

clumps trap fine particulate matter, carried across the desert by wind. These particles include organic matter, litter and spores of mycorrhizal fungi, all having potential benefits, for the plants. The clumps, unlike isolated individuals, trap the particles like a snow fence. Individuals of different species can form clumps more readily than individuals of one species because their roots may not have the same placement and they can, therefore, avoid competition that would occur between conspecifics with the same rooting requirements. The reality of this proposed scenario is being tested by our group on a mined site by planting shrubs in different arrays and assessing their growth rates and the changes in the soil conditions around individual plants as opposed to clumps. To date the clumping proposition seems to be true, but the story is much more complicated than outlined above, and even more complicated than we had ever anticipated. Interestingly, this whole line of research was spawned by the observation that it was much more difficult (some thought impossible) to reclaim mined lands in areas of less than 25 cm of rainfall than in the more humid areas of eastern North America. The mining experience brought out the critical importance of the reaction phase of succession and caused us to delve deeper into processes, casually overlooked previously (Ch. 14).

The importance of vertical architecture to community structure is easily demonstrated by a wide variety of studies for many animal groups, conducted by many researchers. Birds, mammals, insects, spiders and others respond to the presence or absence of structures. As these structures change over time, the composition of the animal community changes. This is not meant to imply that structure is the only driving variable for animal community change, but it is an important one. In our own studies of shrub-dwelling spiders the experimental alteration of the branch density of one shrub species, big sagebrush (*Artemisia tridentata*), altered the spider community composition. Open shrubs (with some branches removed) favored spiders that built webs to snare their prey; closed shrubs (with branches bundled together) favored species that hid and ambushed their prey (Hatley & MacMahon, 1980). Clearly, architecture *per se* is a significant component of the reaction process (Ch. 22).

The results of reaction alter community composition and dynamics. The importance of the process was generally recognized long ago. However, it took information from restoration attempts to illustrate the nature and extent of its importance. Studies to date have been much easier to implement, in an experimental sense, in the context of reclamation work than they have in native communities.

## Ecology and reclamation – the biome-specific imperative

The successional processes described above are clearly pertinent to reclamation efforts. Equally clear is the fact that attempts at reclamation or restoration have allowed ecologists in our group and elsewhere to focus their attention on processes that were not as obvious in native systems as on disturbed sites.

What none of the above discussion addresses is whether or not these

| | DESERT | TUNDRA | GRASSLAND | CONIFEROUS FOREST | DECIDUOUS FOREST | RAIN FOREST |
|---|---|---|---|---|---|---|
| NUDATION | drought | cryo-planation | fire | fire/wind | wind/senescence | senescence/wind |
| MIGRATION | seed reserves | non-seeds | seeds and non-seeds | seeds | seeds | seeds |
| ECESIS | periodic | slow-periodic | moderately fast | mod. fast (variable) | fast | very fast |
| COMPETITION | water | water | water/light | light/water | light | light |
| REACTION | low | moderately low | moderate | moderately high | high | very high |
| STABILIZATION | fast | fast | moderately fast | slow | slow | moderately slow |
| MISCELLANY | no physiog. no species | no physiog. no species | mod. physiog. high species | high physiog. high species | high physiog. high species | high physiog. high species |

*Fig. 15.2.* Processes involved in plant succession vary from community to community in ways that have important implications for restoration, as summarized in this table. Here, biome types are arranged on the basis of the conspicuousness of succession in that biome. Aspects of succession include: nudation (factors most likely to create disturbances in the biome type); migration (processes by which plant migrules or propagules reach the site); ecesis (the speed and reliability with which plants establish from propagules); competition (including the relative importance of water and light as limiting resources); reaction (the degree to which the seral biota alters the site); and stabilization (the rate at which plants typically stabilize or reach "climax", both physiognomically and compositionally). "Miscellany" refers to the degree of physiognomic and species turnover during succession (from MacMahon, 1981).

processes are equally important in all systems, and if they are not, is there
a way to generalize about their varying importance. Over the years, I have
attempted to consider this question. First, I developed a matrix with qualitative statements concerning the importance of the processes in each

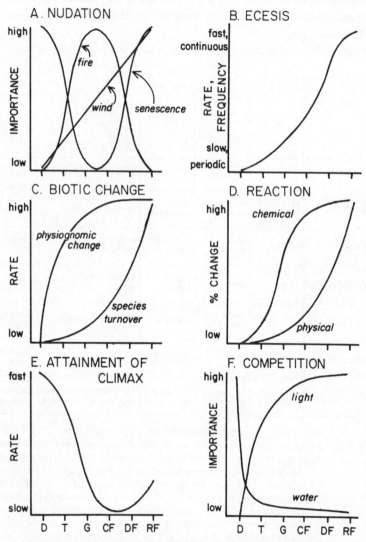

*Fig. 15.3.* A graphical representation of some of the information in
Fig. 15.2. suggests these hypothetical trends for certain successional
processes in a variety of biomes. Biomes included are: Desert (D);
Tundra (T); Grassland (G); Conifer Forest (CF); Deciduous Forest (DF);
and Rain Forest (RF) (from MacMahon, 1981).

biome type (Fig. 15.2). The order of the biome types was fixed by the conspicuousness of succession within that biome. Thus, succession is superficially less obvious in deserts and tundras and most obvious in deciduous and rain forests. These qualitative statements were turned into graphs as a form of hypothesis statement and for ease of interpretation (Fig. 15.3).

This exercise, like ones mentioned above, was prompted by studies of reclamation. The difficulty of reclamation efforts in the west and specific statements from a committee of the National Academy of Sciences and the National Academy of Engineering (National Academy of Science Committee, 1974) suggested that along a gradient of mesic to arid sites, the difficulty of revegetation increased. What was the reason for this?

Perusal of the hypothetical curves suggested that ecesis was slowest and reaction influences were the lowest in arid areas. This makes sense in that there are numerous seeds in desert soils, but the probability of having the proper moisture regimes for germination and establishment is quite low. Rainfall is scant and highly variable in arid areas. Thus, only periodically will conditions be correct for germination. The probability of back-to-back "good" rainfall years is even less. The result is that seedlings often encounter drought and die. Sequences of years of rain sufficient for germination, followed by years wet enough to ensure establishment are rare – about every 40 years or so. This is not the case in forests. Clearly, if you want to revegetate a desert, you must, in the terms we have been using, attend to the ecesis process and be sure that you mimic a reaction product that will allow plants to persist.

The statement in the Academy's report concerning arid areas reflects this. It said, "Revegetation of the areas can probably be accomplished only with major, sustained inputs of water, fertilizer, and management".

Our analyses suggest that skipping the ecesis stage would help restore the community even faster. In essence you put tublings (plants germinated and established in the greenhouse) into the field and rely on their physiological robustness to endure droughts that would kill seedlings. In addition, since we believe that a clumped dispersion pattern would aid in developing beneficial soil properties, plants would be planted in clumped arrays rather than at regular intervals – the current mode on mine sites.

In a mesic area the strategy would be quite different. Here ecesis occurs relatively readily, and it is competition among rapidly growing individuals, the biotic interaction process, that requires the most management attention. We could choose a host of other examples. Suffice it to say that one must look at the specific biome context to determine which of the processes of community development (succession) might be the bottleneck to a successful program.

In a sense we have come full circle. As ecologists, our studies of native communities suggested that reclamation of mined lands might be easier using an ecological approach rather than one based strictly upon the traditional principles and practices of agriculture. The attempt to implement this process forced us to rethink many of our positions and to question the "truth" of some of our postulates. The starkness of the reclamation landscape focused our thoughts, altered our emphases, and alleviated some biases.

We have undergone a scientific succession. We had residual knowledge from our own experiences and the work of colleagues. We were constantly bombarded by migrant ideas at meetings, in the literature, and by verbal communication. Some of these established, others died. The result was a different intellectual milieu – our reaction process. We have cycled from native systems to disturbed systems and back. It is not clear how close we are to that desirable climax state – truth – but I fear that we are still in our pioneer stage and that we will need to cycle between native and reclamation ecosystems for years to come, with both systems helping us to develop an integrated systems approach that would not be possible working with either alone.

### References

Allen, M. F. & MacMahon, J. A. (1984). Reestablishment of endogonaceae on Mount St Helens: survival of residuals. *Mycologia* 76(6), 1031–8.

Anderson, D. C. & MacMahon, J. A. (1985a). The effects of catastrophic ecosystem disturbance: the residual mammals at Mount St. Helens. *Journal of Mammalogy*, 66(3), 581–9.

Anderson, D. C. & MacMahon, J. A. (1985b). Plant succession following the Mount St. Helens volcanic eruption: Facilitation by a burrowing rodent, Thomomys talpoides. *The American Midland Naturalist*, 114(1), 62–9.

Brown, J. H. & Gibson, A. C. (1983). *Biogeography*, St. Louis: Mosby.

Clements, F. E. (1916). *Plant succession: an analysis of the development of vegetation.* Washington: Carnegie Institution Publication 242.

Connell, J. H. & Slatyer, R. O. (1977). Mechanisms of succession in natural communities and their role in community stability and organization. *American Naturalist*, 111, 1119–44.

Dvorak, A. J. (1984). *Ecological studies of disturbed landscapes: A compendium of the results of five years of research aimed at the restoration of disturbed ecosystems.* Springfield, Virginia: National Technical Information Service, United States Department of Energy.

Grime, J. P. (1979). *Plant Strategies and Vegetation Processes.* New York: Wiley.

Hatley, C. L. & MacMahon, J. A. (1980). Spider community organization: Seasonal variation and the role of vegetation architecture. *Environmental Entomology*, 9(5), 632–9.

MacMahon, J. A. (1981). Successional processes: comparisons among biomes with special reference to probable roles of and influences on animals. In *Forest Succession: Concept and Application*, ed. D. C. West, H. H. Shugart & S. S. Levin, pp. 277–304. New York: Springer–Verlag.

National Academy of Science Committee (1974). *Rehabilitation Potential of Western Coal Lands*. Cambridge, Massachusetts: Ballinger.

Naveh, Z. & Lieberman, A. S. (1983). *Landscape Ecology*. New York: Springer–Verlag.

Parmenter, R. R. & MacMahon, J. A. (1983). Factors determining the abundance and distribution of rodents in a shrub–steppe ecosystem: the role of shrubs. *Oecologia*, 59, 145–56.

# V | Restored systems as opportunities for basic research

For once the rearrangements run not against the direction of
nature, but along the same steps that nature herself takes.

Jacob Bronowski, *The Ascent of Man*

One reason for suggesting that the restoration of disturbed ecological communities may have considerable value as a technique
for basic research is the insights that have been achieved as a
result of restoration efforts, even when the research involved has
been largely an after-thought – the by-product of a project
undertaken for more or less "practical" reasons.

Examples of restoration projects providing opportunities
for basic research are not uncommon, and the reader familiar
with restoration will no doubt be able to think of numerous
examples. Here we present four, two of which are drawn from
the experience of the UW Arboretum. While it would be difficult
to claim that these are representative, they do suggest the value
of the restoration effort in dealing with a fairly wide range of
issues. Thus the first, an account of John Aber's studies of
nitrogen cycling in several of the Arboretum's restored forests,
is ecosystem oriented, while Grant Cottam's report of his own
work on the Arboretum prairies, is community oriented.

Both clearly suggest the value of working with communities
that have been set up in such a way as to be in some way out

of place, off-target, or improperly assembled. It is this that produces the stresses that reveal underlying processes. This may limit the success of the restoration in practical terms, but, as Bradshaw pointed out earlier, in restoration ecology outright failures may be more instructive than successes. Rules may be discovered by breaking them!

The short piece about the "sandbox" experiment at Hubbard Brook suggests how studies such as those described by Aber may be extended through work with systems created specifically for experimental purposes. And Tom McNeilly discusses the opportunities disturbed and recovering communities provide for research in evolutionary biology.

*John D. Aber*

Department of Forestry, University of Wisconsin-Madison

# 16 | Restored forests and the identification of critical factors in species–site interactions

Ecological restoration is by necessity a holistic or interdisciplinary challenge. Ecologists require information from many scientific disciplines, in order both to understand how different forms of disturbance alter communities and ecosystems, and to prescribe the best treatment for restoration. At the same time, meeting this challenge presents a unique opportunity for field scientists to develop a broader understanding of how ecological systems work through active participation in the healing process.

The purpose of this chapter is to discuss the kinds of information which might be required to restore systems in different stages of degradation, and to present some examples that demonstrate both the value of restored ecosystems for basic research and the contribution basic research can make to the restoration of ecosystems.

## Context

In a field as new and broad as restoration ecology, the nature of the problem may not at first be clear from a scientific point of view. Just what kinds of questions are raised by communities that have suffered various degrees of disturbance, or have experienced various degrees of recovery or restoration? What kinds of information might these communities provide? Figure 16.1 represents an attempt to distinguish between different types of degraded lands on the basis of the kinds of questions each poses, and the management practices which may be appropriate to each. The three classes are ordered by increasing levels of disturbance (see Ch. 5).

The first and mildest level of disturbance involves disruption or removal of the native plant community, without severe disturbance of the soil. This is labeled "secondary succession" in Figure 16.1. Examples might include reestablishment of forests in recently cutover areas, or restoration of prairies on relatively undisturbed pastures. The different boxes within this stage represent different successional stages within a community type. The arrows indicate several possible paths that succession might take, in keeping with the idea that "succession" need not be a one-directional process and that it can be deflected significantly by disturbance or even chance.

The distinctive feature in this stage of degradation is that the soil is left largely intact. Thus, growing conditions may be close to those in undisturbed communities. In this rather forgiving environment, it may be possible to move directly to establishment of native species selected from mature

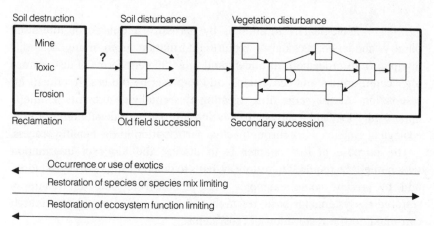

Fig. 16.1.   Restoration projects may be classified on the basis of the severity of the disturbance they are intended to overcome. In general, the kind of questions encountered depends on the degree of disturbance. In its earliest stages, restoration of a profoundly disturbed site (left) involves restoration of function and tends to be ecosystem oriented. Later stages raise community-oriented questions pertaining to population dynamics and species composition and interactions. To some extent, this scheme may provide a way of identifying processes of interest.

"target" communities. This is not to say that this step of controlling the final species mix is simple: the presence of exotic species and native weeds can pose serious problems. Controlling successional change in a mixed community of native and exotic species is often a major task, for example, in prairie restoration in the American Mid-west. The idea expressed in Figure 16.1 is that restoration from this stage raises questions chiefly in the areas of autecology, population biology and species interactions. These questions may arise in the course of selecting species, matching species to sites within the area being restored, or designing management regimes for the developing community. On the other hand, other questions may be of little importance in restoring a community following this stage of disturbance. For example, information on the requirements of particular species for specific soil conditions, such as nutrient and water availability, may not be crucial because the soil environment has not been severely disturbed or degraded.

Information on species–site interactions becomes increasingly important in the second stage of site degradation, when both the vegetation and the soil have been damaged. An example might be an old-field site that has been subjected to plowing and other agricultural purposes for some time. Here, removal of nutrients in crops along with possible erosion and compaction may have lowered both nutrient and water availability. The process of "old-field succession" by which a sequence of species replacements rebuilds both vegetation and soil is one of the classic areas of ecological research.

The restorationist might not be content, however, to let this natural process run its course, but might want to accelerate succession or skip several of its stages in order to produce a mature community faster. In this case, it becomes important to understand both how farming has altered soil conditions and also how the successional process reverses the degradation.

This, in turn, implies an understanding of the growth requirements of the desirable species and the ability of a site to provide those requirements. I feel that ecologists have been severely hampered in this area by a lack of simple, quantitative methods for "seeing" site conditions in a precise and unambiguous way. It is not enough to know, for example, that sugar maple (*Acer saccharum*) tends to grow on heavy soils, and so to infer that it requires both nutrients and water in large amounts. The standard measurements of soil texture, organic matter, pH, etc., are not directly important to the physiology of the plant. They are, rather, indices to nutrient and water availability – and generally not very accurate ones at that. In the second part of this paper, I will present an example of a simple method for measuring one critical parameter, the availability of nitrogen, in a physiologically meaningful way. It will also be clear from this later discussion that previous restoration efforts, such as those at the University of Wisconsin Arboretum, provide the opportunity to measure a species' response over a much wider range of environmental conditions than in natural systems, where species are often restricted by competition to a narrow set of conditions.

The third category of disturbance includes those areas where the vegetation is completely removed and the soil is converted to a form totally outside the range of natural conditions. An example would be mine tailings where soil organic matter is gone, texture and horizontal structure are completely disrupted, and toxic elements may also be present. Another would be a severely eroded site where only the infertile lower soil horizons remain.

The measurement of environmental conditions and species tolerances and requirements becomes even more crucial in this case. We can no longer assume anything about the suitability of the site for native or exotic species. Trace amounts of toxic substances may be more important than availability of nutrients. In order to deal with such sites, we need to understand how species respond to these new environments, and how they might modify the environment in beneficial ways.

In this most disturbed case, the soil needs to be recreated before the mature plant community will function. This raises a different set of interesting questions as to which soil properties are really the most important in supporting plant growth. The soil alteration itself can be accomplished in one of two ways: the first is by physical, chemical, or mechanical site preparation. This, however, is neither cheap nor elegant; another way is to use plants. The potential problem here is that plants that have unusual physiological mechanisms (for example symbiotic nitrogen fixation) and are therefore more likely to survive and to ameliorate soil conditions are often exotics (see, for example, Ch. 7). The introduction of exotics into restoration projects is likely to be resisted because their removal may pose major problems in the final control of species composition.

Perhaps we can think of the reclamation and restoration of severely damaged sites as a two-step process. In the first, we make use of the vast array of physiological tolerances and mechanisms presented by the plant kingdom to speed the amelioration of site conditions. Two processes in particular, nitrogen fixation and cation pumping, will be discussed below. This is really a form of biotechnology – matching physiology to need without regard for whether the species selected belong to the target community. Once the function (e.g., nitrogen cycling) and structure (e.g., soil organic matter) are restored, then the exotic community can be removed and replaced by a native community. Examples of this approach include reclamation of mine wastes in the humid eastern parts of the USA where black locust (*Robinia psuedoacacia*), a nitrogen fixing legume, and aspen or hybrid poplar (*Populus spp.*), a highly productive cation pump species group, are often planted first. The second stage, however, is generally left to natural succession (see Ch. 7).

In general, restoration needs to be considered in terms of the interaction between species and site. In all but the most forgiving or enriched environments, restoration may be more likely to fail because of inadequate nutrition or soil aeration than because of biological interactions. Thus, restoration draws attention to a serious need to increase our knowledge of the precise environmental requirements of species of interest, as well as our ability to "see" or measure critical environmental factors such as resource availability in soils. Moreover, the soil–plant interaction is a two-way process. Important soil processes such as nutrient release can change as rapidly as species composition. It is a case of a variable approaching a variable, and we need to understand both. Restoration, and studies of restored communities, can help.

### Measurements of resource availability and species requirements in natural and restored communities

Enough of theory! In this section I will present an example from my own research on nitrogen cycling in forest ecosystems that suggests how restored communities provide unique opportunities for studies of ecosystem function, once appropriate measurement techniques are available. Two points will be emphasized: first, it would not have been possible to develop as complete a picture of species-specific nitrogen requirements without the restored communities; and second, the information obtained should prove to be of value in planning future forest restoration projects.

The study has been carried out in two very different types of sites. One is Blackhawk Island, a large (75 ha) island in the Wisconsin River about 70 km north of Madison. Due to a complex early post-glacial history, the island contains a full range of soil textures from sand to silt-clay loam. The plant communities have been largely undisturbed for at least 120 years, and species composition is exactly what would be expected for the different soil types (pines on sandy soils, maples on silt-clay soils, oaks in between, and even hemlocks on organic soils over the sandstone bluffs). It is very nearly a microcosm of the native forests of southern and central Wisconsin.

The second site is the University of Wisconsin Arboretum in Madison, a series of restored communities where species composition is not always in line with resource availability. This disjunction between site and species has led to some difficulties with the restoration, but is crucial to an assessment of species requirements. In mature stands, a species' distribution is restricted by competition to that part of important resource gradients where it grows best. Thus we typically see a fairly narrow range, as is certainly true on Blackhawk Island. It is impossible to know a species' response to a full range of conditions if it can only be found under a very narrow range – hence the value of the restored (and in some instances poorly sited) communities. In the Arboretum, we have discovered nitrogen-rich sites with white pine planted on them, and nitrogen-poor sites with sugar maple.

The method we have used to measure nitrogen availability is called the on-site buried bag incubation technique. It is very simple, involving monthly collection of ten pairs of soil samples in each stand: one of each pair is taken to the lab and chemically extracted to determine the total nitrate ($NO_3^-$) and ammonium ($NH_4^+$) content; the other is placed in a thin polyethylene bag, returned to the soil and allowed to incubate in place for a month. It is then collected and also extracted for total nitrate and ammonium content. The difference between the first and second measurements indicates the amount of nitrogen mineralized or released from the soil during the month of incubation. These values can then be added for successive months to produce an annual total.

While simple, this is certainly not as simple as measuring total soil organic matter or nitrogen at one point in time. The problem with these total content measurements is that about 98 per cent of all soil nitrogen is in organic form and is totally unavailable to plants. It must first be broken down into nitrate or ammonium before it can be used. The rate at which this happens is not related to the total amount in the soil at any one time. Thus by measuring the annual production of inorganic forms of nitrogen we obtain a good measure of availability on a temporal and spatial scale appropriate for the study of landscape units.

How good is this method? Figure 16.2 compares annual nitrogen mineralization with net primary productivity above ground (woody increment plus total litter fall) for three groups of species: pines (*Pinus strobus* and *P. resinosa*), oaks (mainly *Quercus alba*, *Q. borealis* and *Q. velutina*) and maple (mainly *Acer saccharum*). Productivity is important in the restoration context not because it is related to yield of product, but rather because it is related to the potential of the vegetation to ameliorate the site. The accumulation of organic matter above ground can alter microclimate and seedbed conditions. Organic matter in the soil helps adsorb heavy metals, increases water infiltration rates and reduces erosion.

The close relationships between nitrogen availability and productivity here are tighter than previously reported for any other soil-based measure of site quality. From this and other evidence, we infer that nitrogen limitations are important in controlling production in most of these stands, and also that the species groups used to derive these relationships do vary in the efficiency with which they use nitrogen to produce biomass.

One of the surprising results, and an indication of the value of using restored communities for this type of measurement, is the relatively flat

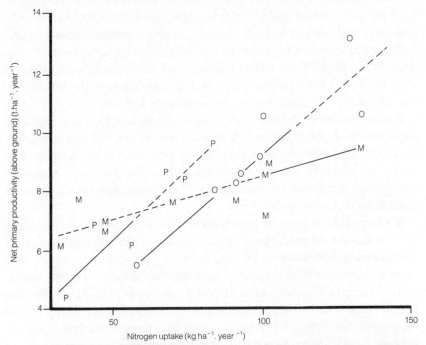

*Fig. 16.2.* Work with restored communities, in which species may have been badly matched to the site, may lead to poor results from a practical point of view, but may also generate information that might not have been obtained in any other way. Here, above ground net primary productivity by three groups of native species (P = pines, O = oaks, M = maples) is regressed against annual nitrogen uptake (mineralization + precipitation inputs − leaching losses). The solid lines show which part of each data set was drawn from native communities. The dashed lines include the range of sites provided by the restored communities in the University of Wisconsin Arboretum (data from Pastor *et al.*, 1982, 1984; Nadelhoffer, Aber & Melillo, 1985; Lennon, Aber, & Melillo, 1985; Aber *et al.*, 1983).

relationship between nitrogen availability and production in stands of sugar maple, a species generally described as "mesic", and hence thought to be both nutrient- and water-demanding. The problem here is in the imprecise nature of the term "mesic". From further work on water use, it seems clear that sugar maple does indeed require water in greater quantities than oak or pine per unit of production. However, it appears to have a relatively low demand for nitrogen, a finding completely consistent with the species' low nitrogen content in leaf litter and other physiological characteristics.

In other words, the measurement of sugar maple responses over the wide range of environmental conditions made possible by the restored communities has enabled us to discriminate between two factors which have usually been lumped together under a single term. The relevance of this to restoration is that we can now distinguish between nutrient and water limitations. Sugar maple, for example, can be planted on nitrogen-poor sites, such as old fields, if the soil is fine textured and high in water-holding capacity. Such a site is currently occupied by sugar maple in the Arboretum, and the maples are surviving, if not particularly thriving.

It may not follow, however, that other species usually associated with sugar maple on the rich sites where it usually occurs will also transfer to water-rich and nitrogen-poor sites. Some, particularly spring ephemeral herbs, are probably very nitrogen demanding. We thus need additional information of this type for those species that we may want to include in the restoration of a sugar maple forest.

It is important to consider the effects of the species on the site, as well as the site control on productivity. Restoration involves the construction of ecosystems that are stable in the long term, so it is necessary to ensure that current production does not come at the expense of future nutrient availability. In connection with John Harper's comments (Ch. 3) on the ecological value of experience in agriculture and forestry, it is interesting that an excellent example of this problem comes from forestry practice. In several regions of the USA, hardwood sites are being converted to pines because of the pines' higher efficiency at using nitrogen to produce wood (e.g., Fig. 16.2). However, this efficiency is achieved through the production by pine species of smaller amounts of leaf (needle) litter with lower nitrogen contents and slower decomposition rates. This, along with a lower total demand for nitrogen, can cause reduced rates of cycling and availability. Forest managers have indeed noted reduced productivity in the second and third growing cycle, following conversion of hardwood sites to pine.

## Other soil modifying processes

Nitrogen use efficiency and its effects on nitrogen cycling and decomposition rates are only one example of how differences between species can alter site conditions. At least two additional physiological processes that can change site quality and that also differ between species groups have been identified, in part as a result of work with disturbed and restored, or at least artificially planted, communities.

The first of these, and probably the most important, is the capacity to fix nitrogen through symbiotic relationships with microorganisms. While the legumes are the most important agricultural group in this regard, several non-leguminous species, particularly in genera of woody plants associated with wetlands (for example, the alders (*Alnus* spp.)) have different but also effective nitrogen-fixing microbial associations. Woody nitrogen fixers are generally found in early successional, disturbed sites where light and water are available in abundance but nitrogen may have been lost, as well as in wetlands where nitrogen availability is low due to anaerobic inhibition of decomposition. The importance of nitrogen-fixing species in altering soil conditions is demonstrated by recent studies comparing forest stands with and without these species (for example, Binkley, Lousier & Cromack, 1984). Soil organic matter and nitrogen content can be increased substantially, as can plant growth and biomass. This effect partly explains the value of nitrogen fixers in reclaiming mine tailings and other severely degraded sites.

Another interesting process is the tendency of certain species to "pump" cations, or move them from lower soil horizons to the soil surface. Recent studies by Alban (1982) contrast the effect of aspen (*Populus spp.*) and white spruce (*Picea glauca*) with red pine (*Pinus resinosa*) and jack pine (*P. banksiana*) on the distribution of cations and pH in the soil. These four species were planted simultaneously on two different sites in northern Minnesota. After about 40 years of growth, both the spruce and the aspen stands have experienced a large removal of cations from the lower soil horizons. These have been incorporated into forest floor and biomass material. The effect of this has been to create a nearly neutral forest floor at the expense of acidifying the lower horizons. The pines have had the opposite effect: the forest floor has been acidified by biological activity, while the lower horizons maintain the high initial pH. If restoration plans called for the establishment of species requiring neutral surface soils, the use of cation pump species might be a more effective way of achieving and maintaining this condition than single or continuous applications of lime.

## Conclusions

Restoration offers the opportunity to increase greatly our understanding of how ecosystems are put together and how they function. To realize this potential, and to derive the kind of information required for successful restorations, we need to know more about how soils and species interact. We need the kind of specific information generated under much more homogeneous conditions by agronomic research. The process of restoration itself can often provide the experimental conditions required to obtain this information. We also need to discriminate between important species groups with very different, useful physiological characteristics. Such information will allow increased precision in prescriptions of treatments to speed the healing process, which is the central concern of restoration ecology.

## References

Aber, J. D., Melillo, J. M., McClaugherty, C. A. & Eshleman, K. N. (1983). Potential sinks for mineralized nitrogen following disturbance in forest ecosystems. In *Environmental Biogeochemistry*, ed. Hallberg. *Ecological Bulletin* (Stockholm), **35**, 179–92.

Alban, D. H. (1982). Effects of nutrient accumulation by aspen, spruce and pine on soil properties. *Soil Science Society of America Journal*, **46**, 853–61.

Binkley, D., Lousier, J. D. & Cromack, K. Jr. (1984). Ecosystem effects of sitka alder in a Douglas-fir plantation. *Forest Science*, **30**, 26–35.

Lennon, J. M., Aber, J. D. & Melillo, J. M. (1985). Primary production and nitrogen allocation of field grown sugar maple in relation to nitrogen availability. *Biogeochemistry*, **1**, 135–54.

Nadelhoffer, K. J., Aber, J. D. & Melillo, J. M. (1985). Fine root production in relation to net primary production along a nitrogen availability gradient in temperate forests: A new hypothesis. *Ecology*, **66**, 1377–90.

Pastor, J., Aber, J. D., McClaugherty, C. A. & Melillo, J. M. (1982). Geology, soils and vegetation of Blackhawk Island, Wisconsin. *The American Midland Naturalist*, **108**, 266–77.

Pastor, J., Aber, J. D., McClaugherty, C. A. & Melillo, J. M. (1984). Aboveground production and N and P cycling along a nitrogen mineralization gradient on Blackhawk Island, Wisconsin. *Ecology*, **65**, 256–68.

*F. H. Bormann, W. B. Bowden,*
*R. S. Pierce, S. P. Hamburg,*
*G. K. Voigt, R. C. Ingersoll, G. E. Likens*

# ☐ The Hubbard Brook sandbox experiment

It is implicit in the idea of the managed restoration of ecosystems that managers and ecologists understand the dynamics of the restoration process itself. One of the greatest difficulties faced by those interested in restoration, especially of long-lived ecosystems like forests, is to obtain definitive knowledge about developmental processes. Only with this knowledge can the manager alter or enhance succession to achieve desired ends.

Obviously, the source of much of this knowledge is basic ecological research. By the same token, restoration and management efforts provide opportunities to test the ideas on which they are based, and frequently lead to new questions of a fundamental nature. Often the two approaches go hand in hand, as illustrated by the following example involving creation of artificial communities to test ideas about nitrogen budgets in forest ecosystems.

Commercial clear–cutting of a northern hardwood forest results in massive removal and accelerated loss of nutrients from the ecosystem. To develop sustained-yield forestry it is necessary to know the rate at which nutrients are restored to the ecosystem. At Hubbard Brook, a hardwood forest in New Hampshire where we have studied ecosystem function systematically and continuously since 1963, we are particularly interested in the restoration of nitrogen, a key element governing growth (see Ch. 16). We have directly measured and modeled nitrogen accumulation in living biomass, dead wood, and the forest floor in a series of stands of various ages since clear-cutting (Bormann & Likens, 1979). We estimate that *net* ecosystem nitrogen accumulation in these forests varies from about 28 kg N/ha-year in 20- to 40-year-old stands to about 2 kg in 140- to 170-year-old stands. Two similar studies have been done in the vicinity of Hubbard Brook on stands that have developed on northern hardwood sites cleared for subsistence agriculture and then abandoned. Both studies indicate rapid *net* ecosystem accumulation of nitrogen despite the fact that previous land use had diminished both biomass and nitrogen stocks even more severely than had been the case in our clear-cut area at Hubbard Brook (Bormann, 1982; Hamburg, 1984).

Where does the added nitrogen come from? Surprisingly, considering the importance of nitrogen as a limiting factor in many ecosystems, little is known about this important question. As a step toward answering it for our system, we developed a mass-balance budget for a 55-year-old stand. Bulk precipitation added 6.5 kg N/ha-year to the stand, while streamwater removed 4 kg N/ha-year. This indicated a *net* ecosystem input rate of 2.5 kg N/ha-year, but living and dead biomass actually accumulated nitrogen at a rate of 16.7 kg N/ha-year, suggesting that there was some additional unmeasured source of nitrogen equivalent to 14.2 kg N/ha-year.

Initially we assumed that the nitrogen input unaccounted for must come from outside the ecosystem by nitrogen fixation and/or dry deposition (Bormann, Likens & Melillo, 1977). Another possibility, however, is that a certain amount of nitrogen is transferred from the mineral soil into living and dead biomass. Such internal transfer (which we had not measured) might be mistaken for *net* ecosystem accumulation, resulting in erroneously high estimates of nitrogen input. This is especially true since the amount of nitrogen in the mineral soil at Hubbard Brook is very large – so large that a transfer of only a small fraction of this mineral soil nitrogen into living and dead biomass might account for a large fraction of what we might otherwise judge to be nitrogen input from outside the ecosystem. Measuring a small loss from the mineral soil is very difficult, however, because the soil profile at Hubbard Brook is stony and extremely variable. As a result, it is virtually impossible to make measurements that are precise enough to detect small losses from the mineral soil.

To gain a better understanding of the potential for *net* ecosystem nitrogen accumulation and nitrogen input rates, we decided to minimize the effect of unknown transfers from mineral soil by looking at much simpler ecosystems where the mineral soil initially contains known, small amounts of nitrogen. Under these conditions *net* ecosystem nitrogen accumulation could be assigned unequivocally to specific inputs – that is, bulk precipitation, dry deposition, and/or biological N-fixation.

We considered two approaches, one taking advantage of suitable natural situations, the other involving construction of an artificial system. In the first case, we proposed to study the revegetation of borrow pits where the soil profile had been removed leaving a sandy substrate extremely low in nitrogen. Under these conditions, we could estimate the initial amounts of nitrogen by analyzing the sand below the rooting depth of current vegetation. Net ecosystem accumulation over time would equal the sum of nitrogen in current vegetation and soil minus the initial soil content. We have now used this approach successfully on moss ecosystems developing on sand in borrow pits.

It is, however, the second approach that is most relevant to the theme of this book. This has involved the creation of monospecific communities in lined boxes filled with sand with known nitrogen content. The idea is that, after an appropriate period of time, we will determine the nitrogen content of the living and dead biomass and the soil in each box. Then, by subtracting the amount of nitrogen present initially, we will be able to estimate *net* ecosystem nitrogen accumulation during the course of the experiment. This will provide an indirect minimum estimate of nitrogen input to the boxes, which can be adjusted by adding annual losses of nitrogen in groundwater and in the form of gas produced by denitrification, both of which we can measure or estimate. Finally, using the Hubbard Brook meteorological monitoring data, we can subtract the input in bulk precipitation from the adjusted yearly input. The difference will provide an estimate of the amount of nitrogen accumulation in these boxes as the combined result of biological fixation, dry deposition, and wind-blown particulate matter.

Overall, our research goal was to obtain accurate information on net ecosystem nitrogen accumulation and its sources. For this purpose, the sandbox approach offered several advantages. For one thing, by isolating the system, it eliminated the major complicating factor of the mineral soil as a source of nitrogen. It should also enable us to distinguish between the various remaining sources – bulk precipitation, biological fixation, and dry deposition – and to assess inputs from these sources with considerable accuracy. Finally, the sandbox experiment should enable us to evaluate the

*Species utilized in sandbox experiment*

---

### Known nitrogen-fixing species

*Alnus glutinosa* (L) Gaertn. – naturalized European species, vigorous nitrogen-fixer. TD.

*Robinia Pseudo-Acacia* L. – nitrogen-fixing species occupying a wide range of sites. TD.

### Species known to be associated with nitrogen accumulation.

*Pinus rigida* Mill. – species of poor, sandy sites. TD and BS.

*Polytrichum juniperinum* – Common moss on the poorest sandy sites. BS.

### Species able to grow on harsh, sandy sites

*Andropogon scoparius* Michx. + *Panicum virgatum* L. – grass mixture widely used in New Hampshire to revegetate sandy sites. TD, M.

*Betula populifolia* Marsh. – early invader of disturbed sandy sites. TD.

*Pinus resinosa* (Ait.) – found on sandy sites in New Hampshire, widely used to revegetate abandoned sand pits. TD, M.

*Prunus pensylvanica* L.f. – vigorous pioneer on nutrient rich disturbed sites, but also common on dry, sandy sites. TD.

Naturally occurring revegetation – whatever comes up. TD and BS.

### Species chosen for other reasons

*Picea rubens* Sarg. – aggressive invader of abandoned pasture at higher elevations in New Hampshire. TD.

*Populus deltoides x P. cv. caudina* (Northest Clone 353) – fast growing commercial tree. TD.

*Rubus idaeus* L. – common on a variety of disturbed sites. TD.

### Control

One sandbox is maintained in a bare condition. TD, M.

---

TD = plot with a dressing of topsoil, BS = plot with sand alone, M = monitored for hydrologic and nutrient export.

role of a number of species in long-term ecosystem nitrogen accumulation. The species chosen and the reasons for their selection are listed in the table above.

While this experiment has been designed primarily in an attempt to identify sources of nitrogen accumulating in ecosystems, this is not our only interest. In fact, we now visualize this long-term study as having the potential to produce basic information on the structure, function, and development of ecosystems that would be of direct use to ecologists and managers alike. Research possibilities include comparisons of the biomass accumulation, nutrient accumulation, soil profile development, faunal population development, and ecosystem structure associated with the various artificially established vegetation types over time. Nutrient-use efficiencies could be studied by comparing the nutrient/carbon relationships of different plant species growing on the same substrate. As an additional example, the effects of different species on hydrologic and nutrient output could be examined. Such studies would not only provide data on the fundamental ecosystem processes associated with nitrogen accumulation, they would also provide a wealth of data on species behavior useful in revegetating sandy sites. In Chapter 16, John Aber illustrates the value of artificially established forest ecosystems in providing opportunities for studies of this kind. The difference here is that this system has been constructed specifically to test certain ideas, and in this sense represents a step closer to a deliberately synthetic approach to ecological research.

Implementation of this experiment got under way in the spring of 1982. We chose a design that would provide well-defined nitrogen sources and sinks and that would have a relatively low long–term cost. The final design called for construction of 12 small boxes (2.5 × 2.5 × 1.5 m) and 6 large boxes (7.5 × 7.5 × 1.5 m) to be filled with locally obtained glacial sands of known, low nitrogen content and planted with species common to dry, sandy sites or chosen for other reasons (see table). All boxes were planted by the spring of 1985. Three of the large pits (or sandboxes as we subsequently called them) were completely lined (four sides and bottom) with an impermeable, Hypalon liner, with outlet drains where hydrologic and nutrient output could be monitored. All the other sandboxes were lined on the sides only, and were open beneath and in contact with the underlying gravelly sand substrate that characterized the site.

In April 1983, we determined the amount of topsoil to be mixed into some of the boxes to promote establishment, the number and size of reference boxes that would remain bare, the species to be tested (including several alternatives in case of failure), the spacing of plants in the boxes, and watering procedures in case of a prolonged drought. Subsequently, topsoil was added to chosen boxes and techniques for sampling the soil in the sandboxes were developed. Soil cores were collected randomly from each sandbox and total starting nitrogen content estimated. Samples were stored for any future analyses we might think important. We also estimated the amount of nitrogen that must accumulate to detect a statistically significant increase. Net ecosystem nitrogen accumulation rates of 10 kg N/ha-year will be detectable after about five years of growth.

Throughout, we took great care to design the experiment for future users. We devised an identification system for the sandboxes that will be used in all measurements of species behavior and subsequent sampling. A computer storage plan for data and a storage plan for soil and plant samples were developed. All notes that document thoughts during planning and construction phases and all data quantifying initial conditions have been gathered into a single notebook, which has been duplicated and stored at Yale University, the North-eastern Forest Experiment Station, and the Institute of Ecosystem Studies.

Height growth of woody plants is measured annually, and observations of plant behavior are recorded. Naturally revegetating plots are mapped annually. The hydrology and chemistry of outflow is measured periodically. Where appropriate, weeds are recorded, pulled and left in the box to retain nutrients.

Our plan is to study the sandboxes for ten to twenty years. Such a long-term experiment will require the dedication and commitment of several key organizations and individuals to ensure continuity of design and purpose. For those planning a long-term study of this kind there are several cautions. One cannot invest too much effort in planning. Gathering materials, soils, and planting materials is a major undertaking. One should strive for simplicity in design, especially in data collection, chemical analyses and frequency of collection. Failure to do so will result in enormous requirements in labor, data and sample storage and analyses, and excessive costs over the course of the experiment.

### References

Bormann, F. H. & Likens, G. E. (1979). *Pattern and Process in a Forested Ecosystem.* New York: Springer–Verlag.

Bormann, F. H., Likens, G. E. & Melillo, J. M. (1977). Nitrogen budget for an aggrading northern hardwood forest ecosystem. *Science,* 196 (4293), 981–3.

Bormann, R. E. (1982). Agricultural disturbance and forest recovery at Mt. Cilley. Thesis. Yale University: PhD Thesis.

Hamburg, S. P. (1984). Organic matter and nitrogen accumulation during 70 years of old field succession in central New Hampshire. Yale University: PhD Thesis.

*Grant Cottam*
Department of Botany, University of Wisconsin-Madison

# 17 | Community dynamics on an artificial prairie

This chapter has three purposes. The first is to relate our experiences at the University of Wisconsin-Madison Arboretum in the restoration of prairies. The second is to draw conclusions from these experiences about the ecology of prairies. The third is to make an assessment of the value of prairie restorations for research to test ideas about the nature of prairies and of ecological processes in general.

Restoration of prairies has been a major part of the Aboretum's restoration program from the beginning (Fig. 17.1), and in general has been its most successful. Of interest here is the fact that two prairies have been restored, using different methods on sites that differ in soil, drainage, exposure and disturbance history. This has provided excellent opportunities to compare and contrast the prairies, in an attempt to assess the importance of various factors influencing their development. Moreover, these are among the oldest restored prairies anywhere. Their restoration has been planned and supervised by scientists. and their development has been closely monitored from the beginning. Although these restorations were not undertaken specifically as experiments, we have learned a great deal from them, both about the ecology of prairies and about prairie restoration and management. Most of the results reported in this chapter are based on a series of inventories of the Arboretum prairies conducted at five-year intervals since about 1950.

Although the two Arboretum prairies are about a kilometer apart and are similar in topography, the contrast between them is striking: except for a small wet area, Curtis Prairie is on an end moraine, with silt loam soil; Greene Prairie is on the site of a glacial lake and has lacustrine soils. These have an average of 50 per cent sand though some areas have as much as 90 per cent sand, and there are lenses of clay that make some of the sandy areas wet (Anderson & Cottam, 1970).

In general, the soil of Curtis Prairie is much better for crops than that of Greene. However, the use of these areas immediately before restoration began was the reverse of what might be expected. The Greene Prairie site had been planted to corn and in a few places furrows are still visible today. The Curtis Prairie site was mostly in permanent pasture when restoration began in 1934. Both sites have rolling topography: Greene is mostly flat with a slight southern exposure, Curtis slopes gently to the north-east. Both sites offer drainage conditions ranging from wet at the lower end to xeric at the upper. Both had been occupied by prairies at the time of settlement nearly a century earlier, but in both cases the original vegetation had been almost entirely eliminated. Greene adjoins a railroad where there were relics of prairie vegetation that provided clues to the composition of the native prairie. And some prairie vegetation persisting on a hectare or so in the wettest part of Curtis was eventually incorporated into the prairie.

Essentially nothing was known about prairie restoration when restoration of Curtis Prairie began in the mid-1930s, and a variety of techniques was used in planting both Arboretum prairies.

In general, preparation of a site for prairie restoration can be carried out in a number of ways. If there is nothing worth preserving on the site, the appropriate technique is to eliminate all vegetation, although only small areas of Curtis and none of Greene were treated this way. This can be done by cultivating several times, applying herbicides, or sterilizing the soil. Soil sterilization destroys much of the microflora and fauna, and this may result in markedly slower growth of the prairie species (see Ch. 14). If there are enough prairie plants on the site to warrant saving them, light disking,

Fig. 17.1.   By the 1950s, a quarter of a century after restoration began on Curtis Prairie, the 24 ha community included more species than any other area of comparable size in Wisconsin, and its rich mixture of species, including more than 300 prairie natives and several dozen weedy exotics, was providing unmatched opportunities for research on the dynamics of prairies (from UW Arboretum files).

raking or burning may be done before planting. Site preparation is more critical if the site is to be seeded than if it is to be planted with transplants. Transplanting of either prairie sods or plants grown in pots or flats usually has a higher success rate, but is much more labor intensive.

The method requiring the least knowledge is the scattering of prairie hay. The technique is simple: just go out and find a prairie of the appropriate type, mow it, gather the hay, and scatter the hay on the prepared restoration site. This method is time-specific and will produce only species that have mature seeds at the time of mowing.

All of these methods work, but the success rate is highly variable and depends on the quantity of weeds present, the amount and timing of precipitation, the way the seeds are stratified, and a number of other variables both known and unknown.

The Arboretum prairies were planted using a variety of techniques, and most of the techniques were used on both prairies. The restoration of Curtis Prairie was the more labor intensive. It was initially planted during the days of the Civilian Conservation Corps, when a large labor force was available, but broadcasting prairie seed after spring burns was continued throughout the 1950s and 1960s. Greene Prairie was planted primarily by one person, Dr Henry C. Greene, who took ten years to get it established. Curtis Prairie was planted first and was the first major restoration attempt ever accomplished. Greene Prairie was planted later, after John Curtis and his students had done a considerable amount of work on prairie communities. As a result, Greene Prairie had the benefit of Dr Curtis's experience of the problems encountered with the various planting techniques, and also the benefit of vastly increased knowledge about the composition of Wisconsin prairie communities and the life histories of the native prairie plants (Curtis, 1956).

Native prairies are usually very heterogeneous, with masses of one species growing together in one place and other species growing together in other places. Why the plants distribute themselves as they do is not easily discerned, so at best a lot of guesswork and intuition goes into the actual planting of the species, and a large amount of background information increases the chance of success. There is room for error, however, because prairie plants generally have a broad range of tolerance. If they are established within the range of their optimum habitat, the species will interact and ultimately sort themselves out into a reasonable approximation of a native prairie. Given the number of higher plant species that should be in a prairie (the average Wisconsin prairie has from 42 to 66 species at a given site (Curtis, 1959)) and the possible interactions among the species, and between the species and various environmental factors, it is not surprising that the degree of precision with which any investigator can locate a particular species is relatively low.

Neither of the prairies was set up as an experiment in how to restore a prairie. There are probably too many variables to permit that kind of experimental design, in any case, and certainly not enough was known then to enable anyone to set up a rigorous experiment. The objective was, more simply, to get the prairies established, and most of the ideas proposed were tried. Much of the record keeping was excellent, especially that of Greene and Theodore Sperry, who was in charge of the Curtis plantings. Some of the later work is not so well documented.

Although the primary purpose for the establishment of the Arboretum prairies was to have samples of prairie for teaching and research, it was expected that the process of establishing the prairies would point up areas where research was required; this did indeed turn out to be the case.

One of the best examples was the research on fire. Curtis Prairie was initially dominated by bluegrasses (*Poa pratensis* and *P. compressa*), which were competing with the planted prairie species. In an effort to find a way to control these and other exotic invaders and to increase the vigor of the prairie species, Curtis & Partch (1948) set up an experiment involving the burning of prairie plots at different times of the year and with different intervals between burns. This was a pioneering effort. No previous studies on the effect of burning on prairies had been published.

In subsequent years other research on the use of fire in prairie reestablishment and maintenance has been carried out, including midsummer burns and the use of properly scheduled fires in the control of weeds (Kline, 1985; Kline & Howell, Ch. 6).

The behavior of clones of *Helianthus laetiflorus* also suggested research opportunities. This sunflower forms dense clones, but only those plants on the perimeter of a clone flower. The species is not only allelopathic – poisoning its neighbors – but also autotoxic – inhibiting its own growth and reproduction (Curtis & Cottam, 1950). Allelopathy is a common phenomenon on the prairie, especially among the Compositae, and the role of this phenomenon in determining the dynamics of species change is a subject ripe for further investigation.

Furthermore, as the prairies developed, numerous instances of unexpected behavior occurred. Several species have exhibited spectacular increases in numbers, sometimes decades after they were first planted. Rattlesnake-master (*Eryngium yuccifolium*) is prevalent only on mesic native prairies and is never dominant, yet it has completely covered Curtis Prairie, in numbers far in excess of its numbers on native prairies. Wild lupin (*Lupinus perennis*) covered large areas in Greene Prairie 40 years ago but now occurs only in scattered patches. White false indigo (*Baptisia leucantha*) did not reproduce at all at first, and the original plants in their rectangular plots could still be identified several decades after planting. Yet now this plant is common over much of the mesic prairie (Sperry, 1984). These instances, and many others, were not anticipated and could not have been predicted from our knowledge of native prairies or the life histories of the species involved. An understanding of the particular combination of climatic events, pest loads, and changes in the environment caused by the prairie itself would contribute much, both to our knowledge of basic prairie ecology and to our ability to restore prairies.

We have been fortunate in having a remarkably good record of the changes that have taken place on the Arboretum prairies. Beginning in 1951 the prairies have been sampled every five years, using a total of about 1300 quadrats laid out on a grid, and using a common base line. It is possible from these data to locate within a few meters the position of every quadrat and to map the prairie communities.

It is even possible to map individual species, but the most useful information has come from mapping the communities. The communities are defined according to Curtis's system, which involves the use of indicator species chosen on the basis of his study of Wisconsin prairies (Curtis, 1956). Curtis established a list of ten indicators for each of five segments of the prairie continuum. The segments constitute a moisture gradient. In surveying a prairie, the number of indicators present from each prairie segment is tallied for each quadrat and the quadrat is assigned to one of the prairie segments on the basis of the number of indicator species it contains. The computation requires the use of weighting values, and the result is a number ranging from 100 to 500, with 100 being the number for quadrats containing only wet prairie indicators, and 500 being the number for quadrats containing only dry prairie indicators. Intermediate quadrats usually contain indicators from more than one prairie segment, although it would be possible to have a quadrat containing only indicators for dry-mesic, mesic, or wet-mesic prairies.

The results that follow are based on syntheses of five surveys made since 1951. Two surveys are omitted because they were based on a closed list of species. Most of the data presented are taken from a thesis by Thomas Blewett, who performed the most exhaustive analysis of the data yet to be done; the results of his investigations are available for anyone interested in the details (Blewett, 1981).

Results for Greene Prairie are shown in Figure 17.2. Two characteristics are important: the first is that a certain amount of shifting in the distribution of community types does occur during the 25 years between the first and last survey, as introduced species find their way to optimum sites; the second, however, is that there is also a great deal of shifting that cannot be accounted for in this way. In fact, the various communities do not have static borders, but appear to change their location with every survey in an amoeba-like movement that seems to be a response to the short-term climatic events of the years immediately preceding the survey.

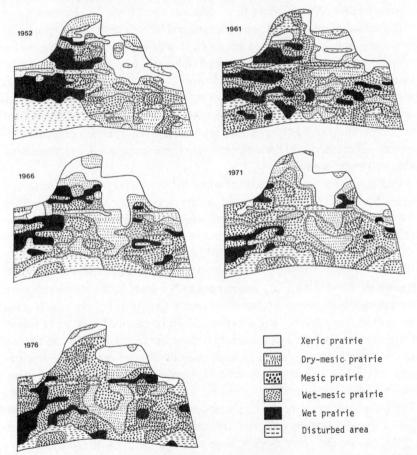

Xeric prairie
Dry-mesic prairie
Mesic prairie
Wet-mesic prairie
Wet prairie
Disturbed area

*Fig. 17.2.* Changes in the distribution of community types on the UW Arboretum's restored prairies continued over the 25-year survey period and were only partly due to consistent shifts of planted species toward optimum sites. Much of this change apparently reflects short-term variations in climate and may be expected to continue indefinitely. The results for Greene Prairie are shown.

If one compares only the earliest and latest surveys, the most noticeable change is the reduction in the areas classified as disturbed – those areas in which a majority of the quadrats had no indicator species at all. On Curtis Prairie there is a phenomenal increase in the area occupied by mesic prairie, and on Greene Prairie it is the wet-mesic segment that shows the greatest increase. This comparison of the first and last surveys, however, ignores the most interesting aspect of the community dynamics. The fact that the boundaries between the prairie segments are in a constant state of flux means that a survey conducted in a given year can really be used only as a point in time. At any other time, different results will be obtained.

Individual prairie species were also studied by examining their change in frequency from one survey to the next. It was expected, since this was a restored prairie and could be presumed to be unstable, that there would be marked changes in species frequency over the period. The most successful species would be expected to increase in frequency with each survey; other species would be expected to decrease. An examination of the data resulted in classification of the species into four categories. The two expected categories – increasers and decreasers – were there as expected, but there were also two more groups.

One group showed very little change, neither increasing nor decreasing over the 25-year period. The last group was composed of species that fluctuated from one survey to the next, which we termed "sensitive" species. In general, the increasers were the major prairie species and the decreasers were weeds or native plants with weedy tendencies. The species that fell into the "no-change" category were a mixed group. Some of them were prairie species that had apparently achieved stability early in the development of the prairies, and some were persistent weeds. On Curtis Prairie, which is ten years older than Greene, and therefore might be expected to be more stable, there are more no-change species than there are increasers. And on Greene there are four times as many increasers as no-change species. The fluctuating species are the most difficult to interpret. The seeds of one of them, sweet clover (*Melilotus alba*), a biennial, are stimulated to germinate by fire. Thus, the fire history of the two years prior to the survey date could have had a marked influence on the quantity of sweet clover observed (Kline, 1985). As for the other species, their behavior remains unexplained.

Species behavior was also studied by means of species ordinations. The ordination method used was that described by Bray & Curtis (1957) with selected endpoints (Fig. 17.3). The endpoints of the first axis were selected from Curtis's list of indicator species – one from the wet prairie indicators and one from the dry prairie indicators. This axis represented a moisture gradient. The second axis was a gradient of responses to the restoration, with an increaser at one endpoint and a decreaser at the other. The position of a species on either axis reflects the frequency with which it was found to be associated with the selected indicator species.

Ordinations were run for each prairie for each of the five surveys. Since the endpoints of the ordinations were the same for each survey, any change in the location of a species in successive surveys would indicate how the species was responding to the changing conditions. Thus it was possible to construct a time series ordination on which the locations of the species in each successive survey were plotted and connected with a line, giving a trajectory defining the movement of the species.

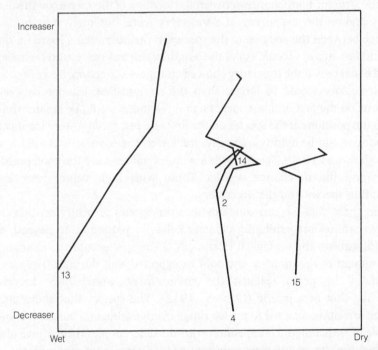

*Fig. 17.3.* Introduction of plants in prairie restoration at UW Arboretum set the stage for a long-term study of movement of species within the developing community. The behavior of species differed widely. Here, for example, species 13 shows the pattern of a wide-amplitude prairie species, moving from the wet, decreaser corner of the ordination in an almost straight line toward the wet-mesic portion of the increaser end of the ordination. Species 4 is a weedy prairie plant that moved from the decreaser end to the center of the ordination. Species 14 is a wet-mesic increaser that moved from the center of the ordination to the top. Species 2 and 15, both weeds, show erratic fluctuations (redrawn from Blewett, 1981).

It was expected that the good prairie species – the increasers – would exhibit fairly smooth trajectories toward the increaser end of the ordination, and this was indeed the case. There was a slight indication that these species also spread out on the moisture axis, but this trend was so slight that it could not be substantiated. The decreaser species showed more erratic trajectories, but they also tended to move toward the increaser end of the ordination and away from the decreaser end. This is due to the fact that, as the prairies become increasingly integrated, the decreaser end of the ordination is tending to disappear, and those decreasers still remaining are growing in close association with prairie plants.

An index was calculated that compared the total length of each trajectory with the straight-line distance between the location of the species on the first survey and on the last survey. If a trajectory were absolutely straight, the distance between the position of the species in the ordinations based on the first and last surveys would equal the length of the trajectory, and the index would equal one. If the trajectory showed changes of direction, however, the total trajectory would be longer than the straight-line distance between locations on the first and last surveys, and the index would be greater than one. If the positions of the species on the first and last surveys were identical, the index would be infinity. Actually, the lowest index value was 1.01 and the highest was 10.86. Species with low index values were the good prairie plants and the no-change species. Those with high values were the fluctuating species and the decreasers.

From these data we can conclude that the species presently existing on the two prairies are exhibiting different behavior patterns that depend, at least in part, on the ecological tolerances of the species and the changing environment of the prairies. It would be expected that during 50 years of occupancy by prairie plants, the environment would have become more like that of a prairie (Gleason, 1912). The species that show little change are either adapted to a wide range of environmental conditions and capable of withstanding competition with the different plants that have also occupied the site, or are very persistent plants living in an environment to which they are no longer well adapted. The sensitive species display the opposite behavior: they appear to be able to become established quickly, and to disappear equally quickly when short-term environmental conditions change. They respond more to climatic variations and to chance disturbance events than to the changing composition of the prairie. The decreasers are slowly disappearing, and perhaps the most remarkable aspect of their behavior is the long period of time they can continue to exist.

The two prairies are very different, and an examination of these differences is informative. Both prairies encompass the range from wet to dry, although the dry part of Curtis is artificial – the result of the deposition of about 0.5 m of coarse crushed limestone on the original silt-loam soil. This attempt to establish a dry prairie was only partially successful, because most prairie species have root systems that go much deeper than the limestone, and many non-dry species have become established on the site and are doing well.

Both prairies are good prairies if one uses the criterion of success of prairies species. Curtis Prairie is on better soil, and its use as a pasture immediately prior to restoration meant that it had a large number of perennial weeds. These have persisted to the present time. Furthermore, this prairie was planted to rectangular blocks of single species and, as expected, it has taken some time for the species to reassemble themselves into the multi-species communities that characterize native prairies. Because many of the blocks were planted with sods lifted from native prairie that contained a variety of species in addition to the one intended, the monotypic nature of the plots was never perfect, but in many instances it is still possible to identify the original blocks (Sperry, 1984).

Greene Prairie was planted with a much greater degree of sophistication. Greene collected the seeds and sods himself and took great care in matching the propagules with the physical environment of the prairie. As a result, less self-sorting has been necessary, and there has been, on the whole, less movement of the species on Greene than on Curtis Prairie. The result is a more spectacular, more natural looking prairie.

The major reason for the difference between the two prairies, however, is the substrate. Curtis Prairie is on the richer site, and in prairie restoration this is likely to be a disadvantage because rich sites are much more likely to be inhabited by weeds. Most of the weeds are adapted to agriculture and normally grow under conditions that will support crops. On poor sites, the native prairie species appear to have more of a competitive advantage, although the general vigor of the prairie plants in the absence of competition from weeds is better on richer sites.

Blewett's primary concern when he analyzed the data was with the autecology of the species. In addition to the work already reported, he analyzed more than 600 soil samples for texture (per cent of sand, silt and clay), for the three major nutrients (nitrogen, phosphorus and potassium), and for organic matter. He also measured three different aspects of soil color, recorded the depth of the $A_1$ horizon and made an assessment of drainage. He presented mean values for all the major species for each of the above characters. The only significant positive correlations he found between the species and the factors he measured were with those that relate to soil moisture and to pH. The main nutrient values for those species that occur on both prairies are markedly different on each prairie, showing more than anything else that the two prairies do differ in soil texture and in quantities of the major nutrients. One of the hypotheses Blewett wanted to test was that the increaser species have broad amplitudes of tolerance for the characters he measured, and this is indeed the case. Perhaps the most outstanding example of this is big bluestem grass (*Andropogon gerardi*), which Curtis (1959) found on more than 90 per cent of the dry, dry-mesic, mesic, and wet-mesic prairies, and on 64 per cent of the wet prairies he studied. This characteristic of the major prairie species is one of the reasons prairie restoration is usually successful.

What do the data tell us about succession on prairies? Essentially nothing. Other data, from ten permanent quadrats that were plowed but not planted, in conjunction with some early experiments on the effect of cutting prairie plants, show that major changes – the expected sequence of annual weeds, biennial weeds and longer-lived perennials – occurred during the first five years. After that time essentially none of the species that had colonized the original plowed field remained. The results reported above are based primarily on data collected ten or more years after the initial attempts at prairie reestablishment, so that any early successional stages would have been complete before that time. The data do show shifts, but they are more likely to be due to climatic fluctuations than to succession. In any case, if succession is occurring, its influence is masked or inextricably combined with fluctuations due to other factors.

The third purpose of this paper is to present my conception of the utility of restoration as a technique or opportunity for conducting basic ecological research. There is no question that the process of restoring a prairie raises many questions that are excellent topics for research, and several of these have already been mentioned. A prairie in the process of restoration also offers opportunities for manipulations that will shed some light on ecological processes. Kline & Howell, (Ch. 6), for example, raise the question of artificial buffalo wallows to perpetuate species that require a particular type of disturbance. Transplantation experiments on prairies in the process of restoration should answer questions about the specific establishment requirements of some species and could also be used to study ecotypes. Conceivably, some light could be shed on the question of initial floristics versus relay floristics (Egler, 1954; see also Ch. 6). Some of these questions could be incorporated into the original restoration plan and initiated at the time the initial plantings are made.

The idea of setting up a restoration to answer specific hypotheses, however, presents many difficulties. One thing the Arboretum experience has taught us is that the unexpected is to be expected, and that we still possess far too little knowledge to be able to anticipate many of the most important developments that will occur. Given knowledge of the life histories of all the species involved, and given detailed knowledge of the major factors of the physical environment, this type of planning and execution may perhaps be accomplished. Given our present knowledge, any grand plan of starting a restoration with the expectation of achieving answers to specific, basic ecological questions is likely to result in disappointment and be a waste of time.

### Conclusion

In prairie restoration there is no substitute for knowledge – especially knowledge of the physical factors of the site, including soil texture, nutrients, and, perhaps most important, soil moisture conditions. Knowing these factors makes it possible to match the site with that of a natural prairie in the same region and to use the natural prairie as a model to design the planting program. It is also useful to know the ecological life histories of the species involved, although prairie species appear to be particularly robust and are capable of growing, at least initially, on sites that are quite different from their optimum habitat. After this comes the problem of site preparation, especially the elimination or reduction of the weed population and the provision of an appropriate seedbed for those species with known seedbed requirements. The provision of a cover crop may often be desirable.

Once the prairie is planted, the most important ingredient is time. With enough labor, it is possible to speed the establishment process by weeding the plantings but this is possible only with small prairies and a big budget. The two Arboretum prairies have a combined area of about 40 ha, and weeding these prairies was out of the question. The knowledge of seedling taxonomy necessary to be a good weeder of a brand new prairie is so great that only an exceptional taxonomist possesses this ability. Good management will, however, reduce the weed problem, but only over a period of time. Curtis Prairie, for example, still has a large complement of weedy species 50 years after its restoration began (Cottam & Wilson, 1966).

A final ingredient in prairie restoration is luck. The most careful plans can go completely awry if the year of the planting turns out to have a prolonged drought. Germination is a problem. Various experiments on methods of stratification may be of some help in predicting the relative density of seedlings, but even under standardized conditions germination percentages may range from near zero to almost 100 per cent. Our most notable failure in this regard was the time we used Indian grass (*Sorghastrum nutans*) as a cover crop. We expected 5 per cent germination and got 95 per cent, and the Indian grass completely dominated the resulting prairie!

Prairie restoration is an exciting and rewarding enterprise. It is full of surprises, fantastic successes, and abysmal failures. You learn a lot – usually more about what not to do than what to do. Success is seldom high, but prairie plants are resilient, and even a poor beginning will in time result in a beautiful prairie.

Prairie restoration is also a humbling experience. You find out how little you really know and how intricately interrelated the species and their physical environment really are.

### References

Anderson, M. R. & Cottam, G. (1970). Vegetational change on the Greene Prairie in relation to soil characteristics. In *Proceedings of a symposium on prairie and prairie restoration*, pp. 42–7. Knox College Field Station Pub. 3.

Blewett, T. J. (1981). An ordination study of plant species ecology in the Arboretum prairies. Wisconsin: University of Wisconsin PhD thesis.

Bray, J. R. & Curtis, J. T. (1957). An ordination of the upland forest communities of southern Wisconsin. *Ecology Monographs*, **27**, 325–49.

Cottam, G. & Wilson, H. C. (1966). Community dynamics on an artificial prairie. *Ecology*, **47**, 88–96.

Curtis, J. T. (1956). A prairie continuum in Wisconsin. *Ecology*, **36**, 558–66.

Curtis, J. T. (1959). *The Vegetation of Wisconsin: An Ordination of Plant Communities*. Madison: University of Wisconsin Press.

Curtis, J. T. & Cottam, G. (1950). Antibiotic and autotoxic effects in prairie sunflower. *Bulletin of the Torrey Botany Club*, **77**, 187–91.

Curtis, J. T. & Partch, M. L. (1948). Effect of fire on the competition between blue grass and certain prairie plants. *The American Midland Naturalist*, **39**, 437–43.

Egler, F. E. (1954). Vegetation science concepts. I. Initial floristic composition – a factor in old field vegetation development. *Vegetatio*, **4**, 412–17.

Gleason, H. A. (1912). An isolated prairie grove and its photogeographical significance. *Botany Gazette*, **53**, 38–49.

Kline, V. M. (1985). Response of sweet clover (*Melilotus alba*) and associated prairie vegetation to seven experimental burning and mowing regimes. In *Proceedings of the 9th North American Prairie Conference*. Moorhead, Minnesota.

Sperry, T. M. (1984). Analysis of the University of Wisconsin-Madison prairie restoration project. In *Proceedings of the 8th North American Prairie Conference*, pp. 140–7.

*Tom McNeilly*

Department of Botany University of Liverpool

# 18 | Evolutionary lessons from degraded ecosystems

It is impossible to pinpoint the time in prehistory when human activities began to effect the sorts of changes in ecosystems that might be termed degradation. There is little doubt, however, that human beings were altering the landscape in a major way, principally through agricultural activities, as early as 11 000 to 9000 BC in the Middle East, about 7000 BC in western China, and between 7000 and 4000 BC in Central America. In some cases the results of this early activity were profound enough for the effects to be still discernible. Similarly, with the development and spread of knowledge of copper metallurgy during the Bronze Age, and further developments in the Iron Age, human activities came increasingly to cause localized and curious degradation of ecosystems, the effects of which may still be seen in parts of Europe – such as at the famous mines of Rio Tinto in south-western Spain.

The most obvious examples of degraded ecosystems are, indeed, the products of past and present industrial activity, and none are more spectacular than the vast derelict – and usually bare – waste products of copper, lead and zinc mining and smelting enterprises. Of course, the influence of activities of this kind on the landscape has grown rapidly in recent times. While copper and zinc have been mined since the Bronze Age and lead since the Roman period, by far the greatest impact resulting from mining of these heavy metals has occurred since the mid-eighteenth century in Europe (Fig. 18.1) and elsewhere, and extensive mining and smelting of cadmium, aluminum and nickel have taken place only during this century.

In evolutionary terms the degraded lands resulting from this activity are therefore of very recent origin indeed. Yet they represent major changes in the environment, analogous to the creation of islands, characterized by extremes of temperature, toxicity, exposure, and drought, as well as low nutrient status and changes in the physical properties of the soil (see Ch. 5).

*Fig. 18.1.* Heavy metal-tolerant populations colonizing mine wastes are providing valuable insights into the patterns and mechanisms of evolutionary change. Shown here is a site at Cwm Symlog, Dyfed, mid-Wales, where mining occurred from about 1700 until 1901. Tolerant plants now growing on the site include populations of tufted hairgrass (*Deschampsia caespitosa*), red fescue-grass (*Festuca rubra*), bentgrass (*Agrostis capillaris*) and a species of catchfly (*Silene maritima*) (photograph courtesy of A. J. Tollitt).

Above all, however, it is soil toxicity that is the overriding characteristic because lead, copper and zinc are so extremely toxic to plants. At the same time, the effects are nearly always limited spatially, and confined to a specific terrestrial source – the mine itself – or aerial source – the stacks at smelters. The degraded areas contrast dramatically with the surrounding normal ecosystems, and often have sharp boundaries.

In the past, similar degradation has been caused by sulfur dioxide whenever coal has been burned or sulfide ores smelted. This has again been associated with specific sources, usually in urban areas, but sometimes in rural sites, such as Sudbury in Canada. Because in this case the major causal factor is a mobile gas, the boundaries of these areas are much more diffuse than those in areas characterized by heavy metal contamination from the mine itself.

There is no doubt that heavy metals at the relatively low concentrations of a few µg/g, and sulfur dioxide at even lower concentrations, have debilitating effects upon plant growth, and can cause death at slightly higher concentrations (Bradshaw & McNeilly, 1981). However, a recurrent phenomenon on sites affected by heavy metals is the occurrence of at least some plants. Similarly, even in areas with considerably elevated levels of sulfur dioxide, some plants do survive.

This immediately suggests that these plants are the product of natural selection, and that these sites offer unusual opportunities for studies of this process and of evolution generally. As long ago as 1934, it was clear that plants of catchfly (*Silene dioica*), growing on an abandoned copper mine in Silesia, were the product of natural selection. Seed from plants growing on the mine produced seedlings that survived on copper-contaminated soil, while seedlings from plants growing on normal soil failed to survive on copper-contaminated soil (Prat, 1934). Similarly, though somewhat later, Dunn (1959) reported that populations of lupin (*Lupinus* sp.) that occurred in the urban areas of Los Angeles were markedly more resistant to smog than populations from coastal areas with clean air. The data of both Prat and Dunn were largely ignored for some 20 years after their publication. However, the message that clearly arises from them is the same: new habitats may well mean new evolution.

It is perhaps worth restating the basic requirements for evolutionary change and relating them to what occurs in degraded ecosystems. Firstly, there must be selection. Degraded ecosystems obviously provide extreme environments for plant growth. In some cases degradation may be so extreme as to devastate the original ecosystem completely. However, these extreme conditions provide extreme selection pressures that might be expected to promote relatively rapid and dramatic changes in the population of whatever species inhabit them. Secondly, for such selection to be successful, there must be variation among the individuals making up the affected populations. In an old, long-undisturbed environment there may be little free variation present in the existing populations because variation has largely been fixed by selection. There may, however, be variation at low frequencies, which changed selection pressures resulting from environmental degradation can exploit. Provided such variation is present, strong selection should lead quickly to the evolution of resistant individuals, as illustrated by the work of Prat (1934) and Dunn (1959). Finally, if, and only if, such variation is genetically based, will the advantageous characteristics be passed on to their offspring to produce permanent changes in the genetic structure of the population. The evidence, as we shall see, is that such variation is sometimes present and is very clearly genetically based. As a result, evolutionary change does indeed occur. Let us examine in detail the remarkable evidence that studies of degraded ecosystems, exemplified mainly by those affected by heavy metals, have provided about evolution.

## Studies of degraded ecosystems

The devastation presented by old metal mine workings in Europe can be awe-inspiring. As a result of inefficient extraction methods, the concentrations of metals remaining in the waste material are often so high as to make reworking profitable. For example, concentrations of zinc may be as high as 14 per cent, lead may be as high as 4 per cent, and copper may be as high as 0.8 per cent (Bradshaw & McNeilly, 1981). The effects of aerially deposited metals is equally devastating, as can be seen in the vicinity of a copper refinery near Liverpool in the United Kingdom (Wu, Bradshaw & Thurman, 1975). Equally dramatic are the effects of the large nickel refinery at Sudbury, Ontario, on surrounding ecosystems, as recorded by Winterhalder (1978) and Cox & Hutchinson (1980).

It has frequently been shown (Jowett, 1964; McNeilly & Bradshaw, 1968; Baker, 1978) that the toxicity of these metals can readily be studied in experiments with solution cultures. These experiments show that the metals primarily affect root growth, concentrations of 0.5 $\mu$g/cm$^3$ copper, 12 $\mu$g/cm$^3$ lead and 7.5 $\mu$g/cm$^3$ zinc typically causing total inhibition. This provides an excellent way of assessing the resistance or tolerance of individual genotypes to these metals, since resistant genotypes can readily produce roots at these same concentrations.

In the UK, severe air pollution began during the industrial revolution, when it became commonplace to burn coal containing as much as 1 to 4 per cent sulfur in small, inefficient furnaces and power stations, and in domestic fires. Ground-level concentrations of sulfur dioxide of 1500 μg/m$^{-3}$ have been commonplace. Average levels of about 500 μg/m$^{-3}$ of this gas reduce the growth of all plants, and kill lichens and conifers. However, certain tolerant individuals can withstand up to five times this concentration. Thus, sulfur dioxide concentrations experienced in the past seem to have promoted evolutionary change just as heavy metals have done.

But what can studies of metal-contaminated sites tell us about existing variation and selective changes in plant populations? Unlike many ecological factors, the factors involved in extreme heavy metal contamination are relatively simple to quantify, and hence it is easy to devise artificial selection programs for them. This, in turn, enables us to examine interrelationships among several phenomena: a severe form of ecosystem degradation; genetic variability within populations; evolution; and the distribution of species.

Surveys of heavy metal-contaminated sites have made it clear that the number of species growing on them is extremely limited. It is clear, therefore, that not all species possess the necessary genetic variability to permit them to colonize such sites (Gartside & McNeilly, 1974a; Walley, Khan & Bradshaw, 1974; Bradshaw, 1984). Thus, although the selective agent is present and is a potent one, it is the availability of appropriate genetic variation that is limiting evolutionary change. This can be generalized, as Bradshaw (1984) has argued, to the conclusion that the availability of appropriate genetic variability is almost certainly an important factor limiting the ecological amplitude of many species.

A further factor emerges from the study – namely the potential of species to evolve characters of ecological significance. Although a particular species may have the general potential to evolve metal tolerance, it may not do so in every situation because the appropriate genetic variability is not present in all populations of the species. This feature has rarely been recognized for natural populations, yet it has recently been clearly shown in bentgrass (*Agrostis capillaris*) (Symeonidis, McNeilly & Bradshaw, 1985). So degraded ecosystems have given us very good evidence on the limits to evolution set by the presence or absence of variability.

In general, it has been considered that evolutionary change is a slow process, and that even micro-differentiation of populations is a slow process. One advantage of mine sites for research on evolutionary processes is that they offer opportunities actually to determine the rate of change under various conditions, since for many sites the age of the mine and the dates of the first and last workings are known. Studies on such sites have made it clear that significant evolutionary change can occur within 100 years or so. However, active creation of a remarkable series of lawns sown with commercial grass seed mixtures at a copper refinery near Liverpool has provided even more detailed evidence. Wu, Bradshaw & Thurman (1975) have shown that significant changes in the genetic characteristics of the plants constituting these lawns must have taken place within four years from sowing. Similar data are available for populations of *Agrostis capillaris* adjacent to a zinc smelter in West Germany, in which changes have been detected after just *12 months* of exposure to zinc emissions from the smelter (Ernst, Verkleij & Vooijs, 1983).

Theory does in fact make it clear that selection pressures that are extreme, as in these examples, can bring about very rapid changes in population structure. However, for a long time the extent to which this would actually occur in the field was not appreciated. The extremely dramatic operation of such selection in the field has now been clearly demonstrated by a study of the population dynamics – and hence details of selection – operating on populations of the grass (*A. canina*) on a small copper mine in mid-Wales (Farrow, 1983; Farrow, McNeilly & Putwain, 1981). The second important lesson provided by degraded ecosystems is that the speed of evolutionary response, even in relatively long-lived species such as plants, can be extremely rapid.

As a consequence of the localized distribution of the waste products of heavy metal mining, and the relatively slow lateral movement of metals in soils, the boundaries between contaminated and uncontaminated substrata are often abrupt. Mined sites thus provide excellent situations in which to assess interactions between gene flow and the process of adaptation to contrasting edaphic conditions – again with a ready means of defining the contrasting types, metal tolerance being easy to assess.

Early work (Jain & Bradshaw, 1966; McNeilly, 1968) established that metal-tolerant individuals are almost entirely confined to heavy metal-contaminated soils, and that distinct populations could exist within as little as a meter of each other, the boundary between them coinciding with the boundary between toxic and normal soils. This is surprising in view of the considerable amount of gene flow that must be occurring as a result of pollen migration across the boundary. Yet boundaries between tolerant and natural populations often remain distinct, even when opportunities for colonization across the boundaries are numerous. At one site examined, for example, there was a considerable amount of disturbance by ants and by grazing sheep, so that potential areas for seedling establishment seemed to be abundant. Yet a very sharp boundary between populations persisted. It has subsequently been established in a series of competition experiments (McNeilly, 1968; Cook, Lefebvre & McNeilly, 1972; Hickey & McNeilly, 1975; Nicholls & McNeilly, 1985) that even if metal-tolerant individuals were to establish on uncontaminated soils, they would rapidly be eliminated as a result of competition from non-tolerant members of the same or other species. This suggests that some penalty is incurred as a consequence of the development of tolerance. In fact, there is evidence that in some populations the competitive ability of tolerant individuals is inversely correlated with their degree of metal tolerance – that is, the more tolerant an individual the poorer its competitive ability is likely to be (Nicholls & McNeilly, 1985).

This, in turn, provides opportunities for studying the ways in which competition and physical factors influence the distribution of a population. Since the extent of colonization of mine sites varies considerably from site to site, tolerant individuals occur in both density-independent or density-dependent situations. In fact, the behavior of populations in competitive situations deliberately contrived for experimental purposes is very close to that predicted by the *r*- or *K*-continuum model of MacArthur & Wilson (1967) and Pianka (1970). Here then, we see a situation in which the distribution of a population is governed to a considerable extent by its competitive exclusion from neighboring habitats.

Studies such as those described above clearly show that metal-tolerant plant populations are the product of natural selection. Related studies involving comparisons of adults and successive seed generations have provided more detailed information about the ways in which selection may influence population structure. The adult individuals are the products of selection throughout the history of that population. In contrast, seed populations produced by adult plants will reveal the extent to which the population might change in response to changed selection pressures, that is, they reveal the potential for change within that population. Comparison of such seed-derived and adult population samples thus provides information about the severity and direction, if any, of selection (Aston & Bradshaw, 1966; McNeilly & Bradshaw, 1968). These data do indeed show that selection pressures for metal tolerance are high, and that they are directional or stabilizing for tolerance.

Comparison of adult and seedling populations may provide valuable information concerning the problems of relatively small, isolated populations – specifically, the potential for gene flow to affect population structure despite selection. Comparison of adult individuals, seedlings derived from wild seed, and seedlings derived from seed artificially produced by those same adults in isolation, will reveal whether or not, and to what extent, gene flow into the adult population is occurring. Such comparisons have been made for populations at the small copper mine at Drws y coed in North Wales (McNeilly, 1968). These revealed a remarkable relationship between gene-flow intensity and selection intensity. High gene flow and high selection together, such as often occur in mine populations, tend to maintain the identity of tolerant populations. High gene flow and weak selection, such as occur on normal soils downwind of the mine, result in a clinal distribution of tolerant individuals on normal soil, a situation in which, except for the high gene flow, they would not occur. A similar situation can be seen on a much greater scale downwind of the very large copper mine complex at Parys Mountain on Anglesey (Bradshaw, 1972).

Perhaps the aspect of the evolutionary process that is most difficult to study is the development of intrinsic isolation between populations. Yet again studies of populations in the extreme environments represented by heavy metal-contaminated soils have provided fascinating information about this process. As we have seen, boundaries between adapted/tolerant and non-adapted/non-tolerant populations are often sharp, distinct populations separated by distances of a meter or less. Clearly, gene flow over such distances is highly probable, and the evidence from seed population studies outlined above clearly shows that it does occur. If the flowering time of such adjacent populations is examined, those plants that occur immediately on either side of the mine/non-mine boundary are often found to have significantly different flowering times. Whereas mine-edge populations flower earlier than non-mine plants, those near the center of the mined area flower at the same time as those entirely outside the mine (McNeilly & Antonovics, 1968; Antonovics, 1966; Nicholls, 1979). These differences have been shown to have persisted over a 20-year period in the field (Antonovics, personal communication). Antonovics (1968) has further shown that the grass populations are the major colonizers of United Kingdom mine wastes, and although almost completely self-incompatible on normal soils, consistently have greater than expected self-fertilization on mine sites. We thus see the maintenance of the integrity of tolerant mine populations being reinforced both by the development of barriers to interbreeding and by increased selfing.

What is the significance of these findings? Since pollen flow is relatively local (Slatkin, 1985), it will be those plants at the boundaries of mines that are the main recipients of pollen, and hence genes from non-adapted, non-tolerant individuals. Their reproductive fitness would thus be markedly reduced compared with those tolerant individuals pollinated predominantly by other tolerant individuals. Any individual in these mine-edge, tolerant populations subjected to reduced gene flow would be at a selective advantage. Reduced interpopulation crossing could most readily be achieved by changed flowering time, or by increased occurrence of self-pollination. Tolerant populations further from mine margins will not – due to the localized pattern of pollen movement – be subjected to the same degree of gene flow, and hence will not be subjected to selection for early flowering, although selfing will still have advantages.

Flowering time has been shown to be a character that can be considerably modified by environmental conditions (Cooper, 1954). The stability of flowering time in mine/pasture boundary populations has been examined by Nicholls (1979). His data clearly show that the tolerant populations growing at the mine boundary had significantly more stable flowering times than either populations entirely outside the mine or those occupying central portions of the mined area. This will tend to minimize the probability of their flowering synchronously with adjacent non-mine plants, a possible consequence of environmental factors. In the face of considerable potential gene flow from adjacent non-mine populations, this will further enhance the fitness of the plants in these mine-edge populations.

Basic to Darwin's arguments in support of the theory of evolution through natural selection is the requirement that features of adaptive significance should be heritable – that is, genetically based. This essential aspect of evolutionary theory has been largely ignored in studies of plant evolution in nature. However, Lawrence (1984) has recently drawn attention to its importance, and it is to be hoped that his paper will stimulate interest in this field of evolutionary study. He drew attention in particular to the relationship, initially predicted by Mather (1953), between the genetic architecture of a character and the type of selection to which that character had been subjected. Fisher (1930) argued that where selection favored a particular homozygote, it might be expected to modify the phenotypic expression of a heterozygote in such a way as to make it approach that of the favored homozygote. Thus, he argued, the favored allele tends to become dominant as a result of the accumulation of modifiers enhancing this effect. In a situation where selection is directional and of such magnitude as to eliminate non-adapted types, selection would thus be expected to bring about evolution of dominance for the favored alleles. In contrast, under stabilizing selection no such accumulation of "one way" modifiers would occur, and dominance would tend to remain ambidirectional. In such a situation much of any genetic variation occurring would be expected to be additive. It follows very elegantly from this that directional dominance for a character reflects a history of strong directional selection.

Degraded ecosystems provide us with excellent situations in which to test these ideas. We have seen that selection pressures promoting and maintaining metal-tolerant populations are strong and directional, selection intensities being of the order of 0.1 to 1 per cent. As with most studies of evolution in natural plant populations, there is scant genetic data upon which to draw. However, what we do have is extremely interesting when viewed in the light of theoretical predictions (Mather, 1960; 1966; 1973).

Working with populations of *Mimulus guttatus* growing on copper mines in the United States, Macnair (1976) has shown clearly that copper tolerance is controlled by a single major gene, and that tolerance is dominant. When we turn to the grass species that have been examined, the picture is much less clear. Zinc tolerance in *A. capillaris* and *Anthoxanthum odoratum* (Gartside & McNeilly, 1974b, c) has a high additive genetic component and is governed by several genes, but is also dominant. In a similar study, Urquhart (1970) showed that lead tolerance in fescue-grass (*Festuca ovina*) was dominant over non-tolerance. These findings are in agreement with Mather's predictions of genetic architecture under conditions of strong directional selection. In an earlier study, however, Wilkins (1960) assessed a larger number of crosses in *F. ovina*, and reported a wide spectrum of genetic architectures for lead tolerance, ranging from dominance of tolerance in some crosses to dominance of non-tolerance in others. The reasons for these differences are unclear, but they may be related to the ages of the various sites in the UK from which the tolerant populations were collected. It may be that on younger sites an insufficient number of generations has occurred to allow the evolution of dominance, even with the intense directional selection for tolerance that is known to occur on these sites.

Finally, and briefly, it must be pointed out that because the products of this evolution are markedly more adapted to metal contaminated soils than the populations of the same species found on normal soils, these populations can have considerable value in ecosystem restoration. (Indeed classical reciprocal transplant experiments to assess adaptation and fitness are impossible in these cases, since non-tolerant individuals simply die on toxic mine soils.) It has been shown quite clearly (Smith & Bradshaw, 1979) that when grown from seed, adapted strains will survive and grow on metal wastes on which non-adapted strains invariably die. As a result of patient work, a number of superior metal-tolerant populations of *F. rubra* and *A. capillaris* have been identified and multiplied for use in stabilization and restoration of mine wastes. One *F. rubra* cultivar ('Merlin') with zinc, lead, and partial copper tolerance is now available commercially, and has been used with considerable success in situations as widely separate as Llanrwst, North Wales, and Palmerton, New Jersey.

From all this, it is clear that studies of degraded ecosystems – in this case mine sites with metal-contaminated soils – have provided remarkable opportunities to advance in many different ways our understanding of evolution through natural selection. Maynard-Smith (1984) states that the theory of evolution by natural selection asserts that any population of entities having the properties of multiplication, heredity and variation will evolve. The entities that occur on metal-contaminated soils have these properties; their progenitors must have had them. Their possession of the fundamental attribute of metal tolerance is crucial but easy to quantify. It is for this reason that the study of evolution of heavy metal tolerance is such a compelling and fruitful one.

Such studies could surely be matched by further studies of evolution on other degraded habitats. Of special interest would be studies of systems that have been subject to various degrees of disturbance. Clearly mine sites represent very severely disturbed systems. The question arises as to how far less severely disturbed systems, such as those often chosen as sites for restoration efforts, may be useful subjects for studies of evolution. The basic principles of Darwin's arguments in favor of evolution through natural selection – availability of appropriate, genetically based variation, and selection, which promotes differential survival of adapted and non-adapted types – will be precisely the same in both cases. At the same time, the *intensity* of selection will be lower on the less severely disturbed sites. As a result, evolution on these sites will occur in the same manner as evolution on more severely disturbed sites, but it will occur more slowly.

Perhaps the best example is that described by Law, Bradshaw & Putwain (1977) for populations of *Poa annua* occupying disturbed sites in urban areas. These populations are quite different from those in pastures, being more precocious in flowering, having a greater proportion of flowering than vegetative tillers, and being poorer competitors than those from more densely populated habitats (McNeilly, 1981; 1984) – a clear case of "*r*-selection". Studies of populations from progressively more stable and well-vegetated areas revealed that the populations did change genetically in relation to this factor, though much more slowly and in a less pronounced manner than they changed in relation to metal pollution.

In general it would seem that natural selection will, and does, inevitably proceed to promote change leading to the evolution of optimally adapted individuals. However, the changes and the processes underlying those changes may be much more subtle than those involved in metal tolerance – witness for instance the remarkable changes shown by Davies & Snaydon (1973) in a population of *Anthoxanthum odoratum* in the park grass plots at Rothamsted. They can nonetheless be followed and unravelled by appropriate observation and experimentation, and can provide further evidence about the meticulous and localized nature of natural selection.

### Conclusions

What, then, are the implications of all these observations to restoration practice? Can we disregard evolution altogether, or can we instead make use of it? Also, in restoring systems how concerned should we be about preserving the integrity of the naturally evolved populations?

Obviously, the use of adapted populations offers one way of coping with the severe conditions often encountered in reclamation projects. Here two methods are commonly used. The wastes may be covered with a nontoxic material that removes the whole problem of toxicity (see Ch. 5). In these situations one can introduce alien species, generally grasses, that are chosen for their agronomic characters and not for any metal tolerance. Alternatively, the waste material can be left, but its fertility, drainage, etc. improved. In this situation the toxicity may remain, in which case suitably adapted, i.e., heavy metal tolerant, populations must be introduced. Again, these may be of alien origin, provided they possess the appropriate adaptations, such as those investigated by Smith & Bradshaw (1979), and provided they are commercially available. Alternatively, because of the speed of evolution and the precision of the adaptation achieved, it may be of value to consider the possibility of using natural populations already beginning to colonize the sites to be restored. On the basis of the experience described above, it would seem probable that these will already have evolved substantial appropriate adaptation. There are now several situations where this is the case, such as where *Cynodon dactylon* is colonizing the Zarwar zinc mine wastes in India (Chaphekar, personal communication). These indigenous populations are generally small and incapable of providing sufficient seed for extensive sowing, but they can quickly be multiplied, often vegetatively.

There is obviously great scope for the use of adapted populations on toxic mine wastes in reclamation programs where the genetic integrity of a natural population may be a minor consideration. The use of such populations may be inappropriate, however, in projects concerned with the actual restoration of natural populations. Generally, even in such situations, little attention is paid to using specifically adapted local populations. Alien material is often chosen merely on the basis of its availability and ecological suitability. This may be quite appropriate, especially since natural selection will then occur to produce closer adaptation to local conditions. So far, however, there has been little investigation of this.

In many situations where complete restoration of the original ecosystem is being attempted, care may be taken to reintroduce the original species. In this case, species continuity may be maintained, but inevitably alien gene pools will be introduced. This will destroy the genetic integrity of the original populations. In common species this may not be a problem, since introduced populations will again be subject to selection of the same sort that promoted the original colonizers, and they may evolve in similar ways. In fact, this evolution may be monitored with advantage, since original material will be available and the demography and time scale of selection, and also its genetic consequences, may be followed in a very rewarding manner.

There are, however, situations in which much more care must be taken. Often commercially available material of a particular species may be grossly different from normal wild material, having been raised, and perhaps specifically bred, for unusual agronomic conditions totally different from those experienced by the wild material. As a result it may not only be very different from the native material in morphology, but also *so* different in ecological adaptation that it does not survive, whereas the wild material does. A good example in Europe is the commercially available form of birdsfoot-trefoil (*Lotus corniculatus*), which is a tall, erect form that evolved in neutral soils under hay meadow conditions and is totally different from the more common wild form, which is adapted to nutrient-poor, acid soils.

Finally, even if great care is taken to use local native seed, there is the possibility that ecologically significant variation occurs even within a particular population. When this is the case it may be necessary to select stock from specific, carefully selected populations even within an area. At first sight this might seem unnecessary because of the flexibility provided by evolution, but it clearly is necessary in some instances. For example, in mineral sand mining in coastal dune systems in Australia, great care is taken to replace completely the original soils and the original plant populations, and there are occasions where seed for several tree and shrub species has to be collected from the appropriate part of the dune system, because the seaward and landward populations are quite different – the seaward being more prostrate, for example – and each is ill adapted to the other's habitat.

In conclusion, the evolution occurring in restoration situations has provided us with a considerable amount of information about evolutionary processes in general. This information clearly has real value for the reclamationist or restorationist. It is important, however, that it be used with understanding and respect, since its misuse could result in harm to the integrity of the very natural populations that are of special interest to the restorationist.

## References

Antonovics, J. (1966). Evolution in adjacent populations. PhD thesis, University of Wales.

Antonovics, J. (1968). Evolution in closely adjacent plant populations. V. Evolution of self–fertility. *Heredity*, **23**, 219–38.

Aston, J. L. & Bradshaw, A. D. (1966). Evolution in closely adjacent populations. II. *Agrostis stolonifera* in maritime habitats. *Heredity*, **21**, 649–64.

Baker, A. J. M. (1978). Ecophysiological aspects of zinc tolerance in *Silene maritima* With. *New Phytologist*, **80**, 635–42.

Bradshaw, A. D. (1972). Some of the evolutionary consequences of being a plant. *Evolutionary Biology*, **5**, 25–47.

Bradshaw, A. D. (1984). The importance of evolutionary ideas in ecology and vice versa. In *Evolutionary Ecology*, ed. B. Shorrocks, pp. 195–202, 23rd Symposium of BES. London: Blackwell.

Bradshaw, A. D. & McNeilly, T. (1981). *Evolution and Pollution*. London: Edward Arnold.

Cook, S. C. A., Lefebvre, C. & McNeilly, T. (1972). Competition between metal tolerant and normal plant populations on normal soil. *Evolution*, **26**, 366–72.

Cooper, J. P. (1954). Studies on growth and development in *Lolium* IV. Genetic control of heading responses in local populations. *Journal of Ecology*, **42**, 521–56.

Cox, R. & Hutchinson, T. C. (1980). Multiple metal tolerance in the grass *Deschampsia caespitosa* (L.) Beauv. from the Sudbury smelting area. *New Phytologist*, **84**, 631–47.

Davies, M. S. & Snaydon, R. W. (1973). Physiological differences among populations of *Anthoxanthum odoratum* L. collected from the Park Grass experiment at Rothamsted. I. Response to calcium. *Journal of Applied Ecology*, **10**, 33–45.

Dunn, D. B. (1959). Some effects of air pollution on *Lupinus* in the Los Angeles area. *Ecology*, **40**, 621–5.

Ernst, W. H. O., Verkleij, J. A. C. & Vooijs, R. (1983). Bioindication of a surplus of heavy metals in terrestrial ecosystems. *Environmental Monitoring and Assessment*, **3**, 297–305.

Farrow, S. J. (1983). Population dynamics and selection in *Agrostis canina* on a small copper mine. PhD thesis, University of Liverpool.

Farrow, S. J., McNeilly, T. & Putwain, P. D. (1981). The dynamics of natural selection for copper tolerance in *Agrostis canina* L. subsp *montana* Hartm. In *Proceedings of the 3rd International Conference, Heavy Metals in the Environment*, pp. 289–95. Amsterdam.

Fisher, R. A. (1930). *The Genetical Theory of Natural Selection*. Oxford: Clarendon.
Gartside, D. W. & McNeilly, T. (1974a). The potential for evolution of heavy metal tolerance in grasses. II. Copper tolerance in normal populations of different plant species. *Heredity*, **32**, 335–48.
Gartside, D. W. & McNeilly, T. (1974b). Genetic studies in heavy metal tolerant plants. I. Genetics of zinc tolerance in *Anthoxanthum odoratum*. *Heredity*, **32**, 287–97.
Gartside, D. W. & McNeilly, T. (1974c). Genetic studies in heavy metal tolerant plants. II. Zinc tolerance in *Agrostis tenuis* sibth. *Heredity*, **33**, 303–8.
Hickey, D. A. & McNeilly, T. (1975). Competition between metal tolerant and normal plant populations: a field experiment on normal soil. *Evolution*, **29**, 458–64.
Jain, S. K. & Bradshaw, A. D. (1966). Evolutionary divergence among adjacent plant populations. *Heredity*, **21**, 407–41.
Jowett, D. (1964). Populations of *Agrostis* spp tolerant of heavy metals. *Nature*, **182**, 816–7.
Law, R., Bradshaw, A. D. & Putwain, P. D. (1977). Life history variation in *Poa annua*. *Evolution*, **31**, 233–46.
Lawrence, M. J. (1984). The genetical analysis of ecological traits. In *Evolutionary Ecology*, 23rd Symposium of BES, ed. B. Shorrocks, pp. 27–63. London: Blackwell.
MacArthur, R. H. & Wilson, E. O. (1967). *The Theory of Island Biogeography*. Princeton: Princeton University Press.
Macnair, M. R. (1976). The genetics of copper tolerance in *Mimulus guttatus*. PhD thesis, University of Liverpool.
Mather, K. (1953). The genetical structure of populations. *Symposium of Society for Experimental Biology*, **7**, 66–95.
Mather, K. (1960). Evolution in polygenic systems. In *Evoluzione e Genetica*, pp. 131–52. Roma: Academia Nazionale de Lincei.
Mather, K. (1966). Variability and selection. *Proceedings of the Royal Society of London*, **164**, 328–40.
Mather, K. (1973). *Genetical Structure of Populations*. London: Chapman & Hall.
Maynard-Smith, J. M. (1984). The population as a unit of selection. In *Evolutionary Ecology*, 23rd Symposium of BES, ed B. Shorrocks, pp. 195–202. London: Blackwell.
McNeilly, T. (1968). Evolution in closely adjacent plant populations. III *Agrostis tenuis* on a small copper mine. *Heredity*, **23**, 99–108.
McNeilly, T. (1981). Ecotypic differentiation in *Poa annua*: inter-population differences in response to competition and cutting. *New Phytologist*, **88**, 539–47.
McNeilly, T. (1984). Ecotypic differentiation in *Poa annua*: within population variation in response to competition and cutting. *New Phytologist*, **96**, 307–16.
McNeilly, T. & Antonovics, J. (1968). Evolution in closely adjacent populations. IV. Barriers to gene flow. *Heredity*, **23**, 205–18.
McNeilly, T. & Bradshaw, A. D. (1968). Evolutionary processes in populations of copper tolerant *Agrostis tenuis* Sibth. *Evolution*, **22**, 108–18.
Nicholls, M. K. (1979). Ecological genetics of copper tolerant *Agrostis tenuis* Sibth. PhD thesis, University of Liverpool.
Nicholls, M. K. & McNeilly, T. (1985). The performance of *Agrostis capillaris* genotypes, differing in copper tolerance, in ryegrass swards on normal soil. *New Phytologist*, **101**, 207–17.
Pianka, E. R. (1970). On r and K selection. *American Naturalist*, **104**, 592–7.

Prat, S. (1934). Die erblichkeit der resistenz gegen Kupfer. *Berichte der Deutschen Botanischen Gesellschaft*, 102, 65–7.

Slatkin, M. (1985). Gene flow in natural populations. *Annual Review of Ecology & Systematics*, 16, 393–40.

Smith, R. A. H. & Bradshaw, A. D. (1979). The use of heavy metal tolerant plant populations for the reclamation of metalliferous wastes. *Journal of Applied Ecology*, 16, 595–612.

Symeonidis, L., McNeilly, T. & Bradshaw, A. D. (1985). Interpopulation variation in tolerance to cadmium, copper, lead, nickel, and zinc in nine populations of *Agrostis capillaris* L. *New Phytologist*, 101, 317–24.

Urquhart, C. (1970). The genetics of lead tolerance in *Festuca ovina*. *Heredity*, 25, 19–33.

Walley, K. A., Khan, M. S. & Bradshaw, A. D. (1974). The potential for evolution of heavy metal tolerance in plants. I. Copper and zinc tolerances in *Agrostis tenuis*. *Heredity*, 32, 309–19.

Wilkins, D. A. (1960). The measurement and genetic analysis of tolerance in *Festuca ovina*. *Report of the Scottish Plant Breeding Station*, 85.

Winterhalder, K. (1978). A historical perspective of mining and reclamation in Sudbury. In *Proceedings of the 3rd Annual Meeting, Canadian Land Reclamation Association*, pp. 1–13. Ontario: Laurentian University.

Wu, L., Bradshaw, A. D. & Thurman, D. A. (1975). The potential for evolution of heavy metal tolerance. III. The rapid evolution of copper tolerance in *Agrostis stolonifera*. *Heredity*, 34, 165–87.

# Doing restoration ecology

Pure scientists and engineers often totally misunderstand each other.

C. P. Snow, *The Two Cultures*

A consistent theme of the preceding chapters is that, whatever its merits in a more fundamental sense, ecological restoration at least provides opportunities for ecological studies that might otherwise be impractical, or even – as Jared Diamond points out – illegal or immoral.

While ecologists may well be eager to carry out experiments involving extensive perturbations, disassemblies and reassemblies of ecological systems, they are naturally constrained in doing so by the same considerations that limit medical experiments with human subjects: in ecology, as in medicine, and in contrast to the physical sciences, the subjects of the research are animate and have a certain irreducible value – some might even say rights. It makes sense, therefore, for the ecologist interested in taking advantage of the opportunities for deeper insight represented by disturbed systems to do as the physician does, and to look to examples that have been disrupted already.

In fact, examples of such systems are numerous, and the experience of a number of scientists suggests that working with them may have a double value, both for science and for the environment. It may also offer unique opportunities not only to do research but to find funding for it. At the same time, it is important in embarking on work of this kind to keep in mind certain considerations that may not apply to research carried out under more traditional conditions. Some of the possibilities, and a few precautions, are described in the following chapters.

*T. F. H. Allen*

Botany Department, University of Wisconsin-Madison

*T. W. Hoekstra*

Rocky Mountain Forest & Range Experimental Station

# 19 | Problems of scaling in restoration ecology: a practical application

Throughout this volume a distinction is made between restoration as a practical matter and restoration ecology, the use of restoration as an acid test of ecology as a predictive science. Our intention, however, is not to insist on the distinction, but to develop the idea of restoration ecology as the basis for a closer relationship between ecological theory and restoration practice. In our view, one way to do this is through the use of hierarchy theory, a new and promising body of ideas that addresses the issue of complexity in multilevel systems. Hierarchy theory has two strengths: it identifies levels explicitly, and so helps ecologists to identify which points of disagreement are genuine and which simply reflect shifts between levels involving mere differences in definition and preferences in point of view; the theory also provides a basis for predictions about changes that occur in complex systems without reducing the system to a series of simple, disconnected systems. Hierarchy theory, through its first strength alone, should be able to contribute to restoration efforts by helping both to define and to devise ways of attacking the problem. Much more significantly, however, restoration ecology (as opposed to ecological restoration in a more general sense) should be able to give hierarchy theory the acid test of its predictive power. If we really understand ecological complexity, then we should be able to predict the outcome of management actions taken in the course of a restoration effort.

In fact, hierarchy theory often leads to predictions that are counter to the conventional wisdom in ecology. What we would like to discuss here is how restoration ecology can help test the theory by testing these predictions. More generally, we would also like to consider how the theory can help resolve issues between the conventional reductionist programs of mainstream ecology and more holistic protocols of the hierarchists. Restoration ecology can do this by presenting the ecologist with a complex system requiring a prescribed action that will produce a particular outcome. The restoration itself is only the final stage, the test, since most of the ecology will already have been done in order to gain the insight that makes the prediction more than an idle hope.

Since plants are the basis of virtually all restoration efforts, it makes sense to begin the discussion with them. Plant population ecology is young, but has made significant progress in the past few decades, a large part of which is summarized in Harper (1977). This success came at the very time when community ecology appeared to be stagnating in a series of methodological arguments about straight gradients, and ecosystem ecology was receiving firm rebukes as a result of the limited success of the large–scale ecosystem simulation efforts of the early 1970s. Now, riding high on their ability to distinguish between neighborhood and population competition, population ecologists dare to take on the upper levels of community- and ecosystem-level ecology, asserting the importance of finding the mechanisms underlying the behavior of these upper level systems. They also suggest that those mechanisms are the aggregation and interlocking of the things that population biologists study, especially competition. Indeed, several of the authors in this volume espouse different versions of that population-centered position.

Hierarchy theorists, in contrast, would suggest that insistent rude reduction to competition or any other asserted rather than discovered principle, is naive and is unlikely to succeed. We may all agree that restoration ecology is a thoroughly worthwhile enterprise and provides one of the few ways to test hypotheses about large ecological systems, but we do not all have to agree what the tests shall be. The important issue here is not that we all concur on the research program for restoration ecology, but rather that restoration ecology itself provides the way of resolving contentious arguments such as holistic versus reductionistic agendas. Better than any other area of ecology, restoration ecology can orchestrate the critical tests of the differences between the reductionist and holistic points of view. It is impossible at this point to be sure, but our surmise is that restoration projects and experiments in restoration ecology will fail to meet predictions based exclusively on lower-level studies of populations and competition. The effective prediction, we suggest, will come from careful reduction from discovered, not asserted principles (Allen, O'Neill & Hoekstra, 1984), and that competition will only occasionally be one of those principles. More important than any of our prejudices, however, is the fact that restoration ecology will be one of the few battlegrounds where any sort of conclusive result can emerge. In restoration ecology the critical predictions can be made and then tested.

### The scale problem

Hierarchy theory organizes the world in a scale-oriented manner. Problems of scale have always been part of ecology, but have received much attention in recent years (Allen & Starr, 1982; Levandowsky & White, 1977). Explicitly, scale-oriented studies have been made of prairies (Allen & Wileyto, 1983), plankton (Allen & Koonce, 1973), forests (Allen & Shugart, 1982), birds (Maurer, 1985), insects (Root, 1974), and food webs and ecosystems (O'Neill *et al.*, 1986). There is much to recommend being explicit about scale in the study of community restoration.

Part of the benefit from a formal statement of scaling problems here comes from the way that restoration specialists know so much about the scale of their research material (for example, a prairie of known size is the object of study and restoration). With such specification of the scale of the enterprise, the ecologist can take full advantage of the body of emerging theory on scale – hierarchy theory. The restoration ecologist is constrained to use prescribed scale, yet the range of scale for these efforts can span orders of magnitude. It is, therefore, crucial that the restorationists understand the ecological effects of scale. Other ecologists are not so constrained and can change the scale of their studies to circumvent problems of working at a particular scale. An ecosystem ecologist, for example, can simplify water and nutrient budgets by expanding a study area until it covers an entire watershed. In general, whatever the most effective – or at least convenient – units of study happen to be, most ecologists are free to choose them. This is not true for the restoration ecologist, who must work with whatever local politics and accidents of ownership provide. Thus problems of scale cannot be circumvented in restoration, so a formal and explicit statement of scale considerations is more important here than it is in some other branches of the discipline.

By the same token, however, restoration projects and experiments provide excellent ways of testing ideas about the importance of scale. Indeed, this may be one of the most valuable contributions of restoration to ecology. Certainly, one of the important sources of confusion in ecology is inexact reckoning of scale, since a change in scale changes the degree of inclusiveness of entities in the system. Accordingly, an inadvertent change in scale has the effect of changing the meaning of common vocabulary. Being free to shave and trim the scale of their study so that it fits conveniently into their preconceptions, mainstream community and ecosystem ecologists are wont to use terms at cross-purposes, and so may well become confused in their attempts to build predictive models. The restorationist, by contrast, has a scale prescribed, and so works in an unambiguously scaled universe with which he must come to terms. Furthermore, though the scale of a given restoration project may be prescribed, the restoration ecologist is in a position to seek out opportunities for experiments on different scales, and even to some extent to create deliberately such opportunities. The forest fragmentation experiments under way in Brazil (Lovejoy *et al.*, 1984), for example, are experiments in scale spanning four orders of magnitude (1 ha to 10000 ha), carried out by deforestation of surrounding areas. Their complement are experiments involving restoration of communities of different sizes. This is very much a part of the restorationist's craft, and the restoration ecologist will certainly take advantage of it. (As an example, consider scaling up Gilpin's experiments with fruit fly communities (Ch. 10), carrying them out, say, in 50-gallon drums.)

### Hierarchy theory

Ecological systems, including those to be restored, are complex. Almost everyone seems to be able to agree on that point, but what is it that we mean when we say this, and to what have we agreed? Too often the complexity of a problem in biology is used first as an excuse to explain why we will never be able to get on top of it, and second as a reason why further research funds are needed. From the point of view of hierarchy theory, however, complexity is not easily ascribed to the system itself, but rather depends on how a description of the system is influenced by the ecologist's relationship to it (Klir, 1985). Thus Allen & Starr (1982) argue that complexity arises when an explanation invokes several levels of organization simultaneously. In other words, when we say that a community is complex, what we mean is that a large number of elements or phenomena defined at a lower level of organization all interact in multifarious ways to produce upper level "community" effects. Complexity presses itself on the observer when he sees, and tries to link, small entities from a low level of organization with large entities from higher levels.

Hierarchy theory does not assert that levels of organization are real in a fundamental sense, but rather that various levels are characterized by the entities that populate them, and that these are defined more or less arbitrarily by the observer. The ranking of levels from high to low does, however, reflect ordering principles related to certain attributes of the entities that characterize the levels. Thus, the higher levels of organization are above the lower levels because their characteristic entities: are the context for; constrain; contain (in nested systems); have less integrity or bond strength than; and, perhaps most important, behave more slowly than entities at lower levels. Since large things generally behave more slowly than physically small things, upper levels tend to be populated by large entities.

As used in ecology, hierarchy theory is very practical; as much as anything it is a theory of observation. Ecological material is so tangible that it takes much restraint to treat field observations as such, rather than as direct knowledge of nature. In ecology, as in all science, there is an observer whose contribution to phenomena cannot be ignored. In physics the role of the observer is well established, and it is recognized that ignoring the observer leads to wrong prediction. The only reason that wrong prediction does not emerge from ecological studies based on the assumption that observation provides ontological truth, is that so few explicit predictions are ever made. In restoration ecology, predictions are at least implied in the goal that elicits management action, so acting on a naive realist philosophy will not do. Implicit in restoration ecology is the idea that it is possible to conduct rigorous ecological experiments which predict, that such and such an action will have such and such an effect. A cornerstone of any such ability to predict with confidence must be a theory of observation.

Observation enters into the question of levels of organization because the pattern of data collection determines which levels will emerge in the study. The lowest level of organization relevant to a given study is determined by the smallest entities that can be discriminated by the measurements (the grain or resolving power of the observation set). The highest level of organization is limited to the one whose entities can only just fit into the entire data set (the extent of the study).

Before we make a scientific observation, we make an arbitrary decision about structure: what is it that constitutes the thing whose state we will measure? Implicit in any such decision is the definition and recognition of surfaces, or the boundaries between entities. Underlying the structural considerations of surfaces, in turn, are processes. In other words, a surface is defined in terms of changes in degrees of interaction. This, in turn, reflects underlying processes.

Consider, for example, a clone of aspen poplars. Tritiated water introduced into one member passes through the entire clone with ease, the time it takes to reach a given stem being simply proportional to its distance from the point of introduction. Here the edge, or surface, of the clone emerges as the place where further movement of radioactivity is slow. The clone is thus defined in these measurements by the process of translocation of water through the underground stems. An entity defined in this way reflects what might be called process closure: the surface is defined as the point beyond which the process attenuates. Across the defining surface of such an entity, points are relatively, but not absolutely, disjunct.

Different observational criteria may be applied to a single entity. In fact, things that can only be observed by one criterion tend not to have much significance. David Bohm made the helpful suggestion that what science seeks is that which persists even when the observational criteria change. For example, a forest can be seen as floristically and physiognomically homogeneous according to biological criteria; but also the forest ameliorates temperature, as can be seen by going in and out of the woods carrying a thermometer. The meandering floristic boundary at the edge of the woods always coincides with the place where the temperature changes, so we are clearly dealing with the same entity. Thus the entity "forest" is robust to the transformation from biological to physical observational criteria.

Robustness to change in observational criteria depends upon the closure of processes like water transport in the aspen clone. Platt (1969) suggests that natural surfaces are places where the closures of several processes mutually reinforce each other. What science seeks is that which is still there even when the scientist observes the system with different apparatus involving new criteria for measurement. The mutual reinforcement of processes and the attenuation of all those processes at the same time and place defines the entities of interest. Only when a large number of processes close together are we dealing with useful entities that display the property of generality and apply in many different circumstances. Tangible entities usually show this coincident closure of processes, but so do critical intangibles such as communities and ecosystems.

## A hierarchical approach to restoration

The prescription of the size of the landscape to be restored determines the lowest frequency process that is relevant to a given restoration effort. If the process has insufficient space or time to behave at its natural rhythm or frequency, then it is not relevant to the restoration. This is, in fact, a general statement of the question of minimum viable populations, since the behavior of single species is only a special case of the behavior of processes in general (see Ch. 20). Restoration is, therefore, limited by the area not so much in quantitative terms (how much area can be restored), but more in qualitative terms (which processes can be included and which cannot). For example, splendid as the University of Wisconsin–Madison Arboretum might be, there is not enough room there to include the process of grazing by buffalo. It is only just possible to include fire as a part of the restoration of the Curtis Prairie. A restored prairie in a residential front yard would probably have to forgo even fire as a restoration process.

Thus, from this point of view, a restoration is not to some ideal pristine system, but only to one whose processes can indeed fit into the available area at hand. This is a limitation to be sure, but not one that should dishearten the restoration ecologist, for restoration in this sense must be seen as area specific, and processes that are too large or too low in frequency to fit the available space are not so much missing as simply irrelevant. Given the area at hand, a herd of buffalo on the Curtis Prairie is an academic point, not a missing part of a "natural" system. An area too small to include the processes of some upper level of organization may be inadequate from an environmental point of view, but for the restoration ecologist it simply poses questions about the level of organization that can indeed fit into the spatial constraints. In this sense, at least, we will no more see the absence of buffalo on Curtis Prairie as a shortcoming than we will regard the biggest tracts of wilderness as inadequate because they are less than global in scale. The essential pragmatism of this viewpoint can provide a clarity of perception that is hard won in ecology outside restoration ecology. Restoration ecology, then, is far more than merely the development of restoration protocols. It has much to offer the rest of ecology in the very fundamental matter of clarity of definition.

Continuing our consideration of what is a natural and proper restoration, we might note that the surfaces and landscape boundaries found in nature are often not the same ones that reflect the limit of an unrestrained process. Trees, for example, create an environment that nurtures saplings and tree seedlings, a process of mutual reinforcement that allows forests to expand.

Even so, there is normally a sharp boundary occurring well before the limit of the fundamental niche of the tree species in question. The natural limit of the community does not have a chance to express itself, being constrained in some other way – for example by natural grass fires, which destroy the invading tree seedlings. Sharp lines of demarcation, namely spatially defined surfaces, are particularly common in human-dominated landscapes (the cows keep eating the seedlings that cross the fence). So natural communities defined on the basis of floristic types located in such landscapes are often locally constrained by landscape processes based on spatial contiguity. Under these conditions, the landscape boundaries hold the community edge away from equilibrium, as is characteristic of upper constraining levels.

Landscapes are not ordered on community principles of floristic or faunal type, but on criteria of spatial arrangement. This, however, does not make landscape ecology intrinsically unnatural, as the grass fire example above suggests. Furthermore, even if a community is situated in an unnatural, human-manipulated landscape, the community is not necessarily unnatural because the distinction between natural and human-dominated ecological systems is commonly a landscape, not a community, consideration. In fact, the difference between landscape criteria and community organizing principles may actually help shield the community from the human intrusion into the landscape. It is common in hierarchies for the level above, with its different surface criteria, to ameliorate the effects of change with respect to lower levels nested within. Thus, the landscape, being in a certain sense the context of the community, acts as a buffer for the community against landscape-based disturbances. A prairie constrained by a fence in a severely modified landscape may not be an unnatural community, because the human-contrived forces are only constraints and are not a part of the network of processes within the community itself. Thus, at least in this sense, there is no need to apologize for the human-intruded context in which the community processes work, since communities ordered by criteria of organism type and population mixes are remarkably independent of the spatial organization of landscapes (see Ch. 21).

It is therefore important for the restorationist or the restoration ecologist to identify whether it is a landscape or a community that is being restored. Vigorous activity on the landscape may have little effect on a community restoration. On the other hand, a community restoration should not be judged a failure just because it appears to have little effect on the landscape in question. One must guard against overestimating the importance of the tangible landscape criteria. Political pressures requiring that restoration be useful and serve the public good will commonly be landscape oriented, because politicians can see a landscape, but might not understand the ordination or cluster analysis that would be necessary to show that a community has indeed been restored. Simultaneous restoration on the basis of several criteria would be ideal, but might well be impossible for purely practical and budgetary reasons. Worse than that, restoration of a landscape may actually result in degradation of communities within it. Conversely, the community-oriented management practice of burning to restore fire-adapted communities might offend the landscape sensibilities of the general public.

Furthermore, important choices for restoration do not always turn on a difference in criteria. Sometimes a difference in scale using just one criterion is enough to produce a conflict. For example, restoration of a rare habitat may require uncoupling community processes, with far-reaching, counter-productive consequences at that upper level. Holding back succession to restore rare pioneer species, for example, is a major perturbation for the community. On the other hand, letting succession proceed may restore community balance while exterminating the last pocket of a rare endemic. The key point is that restoration is scale dependent, so it is important for the restorationist to be explicit as to what is to be restored and the scale of the processes that will run (or apply) to achieve that end. Similar considerations must be taken into account in designing experiments in restoration ecology, and in fact such experiments offer opportunities for testing these ideas and determining their applicability in various communities.

### Conclusion

Perhaps we should be looking for the links among landscape, ecosystem function and community structure so that restoration efforts might benefit from the robustness that we might expect from concurrence of organism patterns and inorganic fluxes. However, such coincidence is so uncommon that restoration ecology will have little to do if it confines itself to such cases. It will be more profitable to accept the differences between organism- and process-oriented accounts, and to be explicit about which of these will be the focus of a restoration effort or an experiment in restoration ecology. Is it landscape, ecosystem function or the community composition that is to be restored and monitored? At best we can hope that the three will have nothing much to do with each other, for there is no reason to suppose that restoration of one will ensure restoration of the others. In fact, much in the same way that conservation of a rare endemic might interfere with community processes, community restoration could easily be carried out at the *expense* of certain ecosystem functions. An important message is that there are very good reasons to be explicit about what it is we wish to restore. The chances are that there are choices to make which, for fiscal reasons alone, not to mention more fundamental ecological reasons, could be mutually exclusive. Let us at least be conscious of the decisions.

Attention to scale is going to be crucial if we are to know what we are doing, either in the practical business of restoration, or in carrying out experiments in restoration ecology. The reason for this careful attention to scale is not just that restorations will fail if we are careless. That would be a pity, but it would not be the most important loss. More significantly, if we allow a cavalier attitude toward scale to interfere with restoration, we will be wasting opportunities to carry out acid tests of whole suites of ecological principles. If we are heedless of the caveats herein, hierarchy theory predicts that almost all our tests of restoration will fail – even those based on essentially correct assumptions. Such assumptions, and the explanations based on them are level-specific, and so explanations that are right at one level are irrelevant at another. In fact they are as wrong at that level as explanations that are downright wrong at every level (Allen & Hoekstra, 1985). In restoration we have a tool to test the validity of basic ecological principles, but failure to scale with precision will blunt even that tool.

### References

Allen, T. F. H. & Koonce, J. F. (1973). Multivariate approaches to algal stratagems and tactics in systems analysis of phytoplankton. *Ecology*, **54**, 1234–46.

Allen, T. F. H. & Hoekstra, T. W. (1985) The instability of primitives and unpredictable complexity. In *Systems Inquiring*, Vol. 1, ed. B. Benathy, pp. 41–4. Proceedings of the 25th annual meeting of S.G.S.R. May 27–31, Los Angeles. Seaside, California: Intersystem.

Allen, T. F. H., O'Neill, R. V. & Hoekstra, T. W. (1984). Interlevel relations in ecological research and management: some working principles from hierarchy theory. *USDA Forest Service General Technical Report RM–110*. Rocky Mountain Forest and Range Experiment Station.

Allen, T. F. H. & Shugart, H. H. (1982). Ordination of simulated forest succession: the relation of dominance to correlation structure. *Vegetatio*, **51**, 141–55.

Allen, T. F. H. & Starr, T. B. (1982). *Hierarchy: Perspectives for Ecological Complexity*. Chicago: University Chicago Press.

Allen, T. F. H. & Wileyto, E. P. (1983). A hierarchical model for the complexity of plant communities. *Journal of Theoretical Biology*, **101**, 529–40.

Harper, J. L. (1977). *Population Biology of Plants*. New York: Academic Press.

Klir, G. (1985). Complexity: some general observations. *Systems Research*, **2**, 131–40.

Levandowsky, M. & White, B. S. (1977). Randomness, time scales, and the evolution of biological communities. *Evolution and Biology*, **10**, 69–161.

Lovejoy, T. E., Rankin, J. M., Bierregaard, R. O. Jr., Emmons, L. H. & Van der Hoort, M. E. (1984). Ecosystem decay of Amazon forest fragments. In *Extinctions*, ed. M. H. Niteck. Chicago: University of Chicago Press.

Maurer, B. A. (1985). Observational scale and community analysis: hierarchical perspectives. *Bulletin of the Ecology Society of America*, **66**, 225.

O'Neill, R. V., DeAngelis, D., Waide, J. B. & Allen, T. F. H. (1986). *A Hierarchical Concept of Ecosystems*. Princeton: Monographs in population biology, 23.

Platt, J. (1969). Theorems on the boundaries in hierarchical systems. In *Hierarchical Structures*, ed. L. L. Whyte, A. G. Wilson & D. Wilson, pp. 201–14. New York: American Elsevier.

Root, R. (1974). Some consequences of ecosystem texture. In *Proceedings of the SIAM-SIMS Conference on Ecosystem Analysis and Prediction*, ed. S. Levin, pp. 83–97. Philadelphia.

*Michael E. Gilpin*

Department of Biology, University of California at San Diego

# 20 | Minimum viable populations: a restoration ecology perspective

MVP is an acronym for "minimum viable population". It suggests the possibility that there is a minimum population size below which population extinction is almost certain to follow. This idea of a threshold is supported in the population biology literature in three different ways. First, MacArthur and Wilson (1967), in their analysis of the colonization of islands by species, found that there was a threshold of about 10 individuals, and that once a population had reached this level it was unlikely to go extinct as a result of birth and death events. Second, animal breeders working with small populations under directional selection have discovered that the inbreeding produced when population sizes fall below about 50 individuals is likely to expose genetically-based deleterious traits that may lead to the extinction of a line, Third, there are so-called Allee Effects in the theory of population dynamics that suggest that some kind of social breakdown is likely to occur below a critical level of population density (Allee *et al.*, 1949). The ability to find a mate is a frequently cited example.

The National Forest Management Act of 1976 has recently led to intense discussion of the MVP concept. Forest managers are required by this act to maintain "minimally viable populations of all vertebrate species on each national forest". That is, they are required to manage habitat – old-growth, successional forest stages, gaps, meadows and so forth – in such a way that each vertebrate species population would remain large enough not to be vulnerable to extinction as a result of chance events. This, of course, includes managing the timing, location and extent of such destructive practices as logging, so as to avoid endangering any local populations of vertebrates.

This seems straightforward enough at first. But, in fact, the idea of MVP turns out to be much more complex than was first appreciated. Gilpin & Soulé (1986) have pointed out that the MVP is actually the multi-dimensional set of thresholds that is revealed by a thorough population vulnerability analysis that takes into consideration genetics, demography, environmental variation, community and distributional ecology, and catastrophes. These authors show that none of these factors can be treated in isolation, for they feed their consequences to each other, producing "vortices" that can lead to the extinction of the species.

Although certainly a complex issue that challenges our scientific know-ledge in the areas of population genetics, population dynamics and community ecology, MVP is nonetheless a practical and pressing issue. How much old-growth forest in Washington and Oregon is required to ensure the persistence of the spotted owl? Should the Yellowstone National Park eco-system be enlarged to provide a sustainable base for its isolated grizzly bear population? If captive black-footed ferret populations can be built up, how should they be reintroduced into a natural landscape consisting of a patch-work of prairie dog colonies?

This last question certainly connects MVP to restoration, for the popu-lation of black-footed ferrets has fallen so low that the continued existence of this species will almost certainly depend on an actual species restoration. This will also be true of the California condor should it respond favorably to captive breeding. The story of the California sea otter, reduced to a single colony of about 18 individuals during the 1920s, also stands as a kind of restoration effort. In fact, although not always treated as such, species reintroduction such as those described by Morton (Ch. 11) frequently include aspects of MVP analysis. Each effort of this kind raises, and provides ways of answering, certain questions: what are the requirements that need to be met in each case to provide sufficient genetic variation, proper sex ratios, the necessary social cohesion, and so forth, to ensure population viability? In fact, restoration is nothing if not a series of tests of MVP.

In this chapter I will not belabor the role of MVP thinking in restoration itself. Rather, I will try to point out that restoration and ecological synthesis in general may actually be one of the best ways of finding out what will happen in the process of ecological degradation and habitat destruction, and that these procedures also provide the best way of identifying the various thresholds that determine the MVP for a species under various conditions.

Consider the case of the spotted owl. It is sitting on what some people believe to be $12 billion worth of old-growth Douglas fir in Oregon and Washington. It is also the kind of standard about which the troops of conservation will rally. Its brush with extinction, should it happen, will be very closely observed, and will be the sort of unfortunate test that will extend our knowledge of the extinction process. Yet the management or mismanagement of endangered species is not the sort of test we need. In the case of the spotted owl and other endangered species, it would obviously be far preferable to explore the MVP question using model systems that will allow the specification of minimal habitat conditions, and that will not provoke the collapse of any naturally occurring population.

As is now traditional in ecology, a systems simulation performed on a digital computer can be a useful way to investigate a system. Unfortunately, there is usually too much latitude in the choice of parameters in such models to stand up to the demands of practical decision–making, especially under conflict. It is thus necessary to use model systems made up of real organisms to identify thresholds for population extinction, but these should be artificial systems, or should involve undesirable populations in the wild – introduced rats, for example – rather than natural populations. Populations of fruit flies offer certain advantages for this purpose since experiments on them can be done cheaply and rapidly (see Ch. 10,). They are also easily repeated. Such experiments are, however, probably best viewed as an extension of computer models, wherein accuracy is sacrificed for generality, rather than as a practical basis for real–world management decisions.

In any event, it is clear that the questions involved are both fundamental and practical in nature. The basic issue is the transition from viability to inviability. What are the biological and ecological attributes of the thresholds, separating these two states? How sudden is the transition across thresholds? Are the transitions different for different processes – for genetic and demographic processes, for example? Under restoration, the questions are reversed. Here one must ask what minima must be reached to change the situation from one of population inviability to viability? When in the restoration process does one introduce the species? How many individuals should be released? What aspects of the landscape and the season of the year need to be considered in choosing the time of release? These questions are especially important for vertebrate species, for which the number of individuals available may be small, and the habitat requirements may be relatively narrow. This may make the task of reintroduction more difficult from a practical point of view. By the same token, however, it means that each restoration attempts offers opportunities to identify the ecological criteria that determine the point at which the viability of a vertebrate species makes a step increase to a higher value.

Most managers of restoration projects would probably desire a margin of safety in releasing vertebrates. Thus, in practice, introductions will usually not match some hypothetical MVP. This complicates interpretation of the results from a theoretical point of view, since it reduces the precision with which successes and failures indicate threshold population levels. Therefore, apart from exceptional cases carried out specifically as MVP experiments, the results will be statistically noisy, and it will be the great numbers of such tests that will be critical in providing information on viability. Where there are a dozen or so endangered vertebrates subject to ecological scrutiny, there will be hundreds of ecological restorations, each involving many tens of vertebrate species. In a fraction of these, the manager will guess wrong, thereby experimentally invalidating at least one of the assumptions on which his or her calculations were based. A database of such introductions, successes and failures, would be an ecological gold mine, not only for ecologists studying MVP but also for those interested in questions of community structure generally.

In fact, the introduction of the species need not even be managed for this approach to work. Most introductions of vertebrates will, in fact, be invasions from neighboring, undisturbed habitats. If properly observed, such "passive restoration" is likely to contribute to the deepening of our knowledge, since the invading species will not follow the timetable of the restoration manager. They will, in fact, invade prematurely or in inadequate numbers. Such invasions will fail. Thus, later, successful invasions will offer opportunities to bracket the conditions of population viability. It needs to be remembered, of course, that these conditions and thresholds are multidimensional. Thus considerable data, properly analyzed, will be needed to reveal them.

The kind of data one would want to put into such a data base would include: (1) the life history properties of the introduced species; (2) its community and food web location; (3) the "state" of the system into which it was introduced including variables measured on many different scales (see Allen & Hoekstra Ch. 19). With such data, one could do the sort of multidimensional data reduction commonly practiced in ecology. Such a reduction could produce, for example, a discriminant function that, as a function of all of these variables, separates the conditions for invasion failure from those of invasion success. Once such a statistical "surface" has been defined for a species, it would provide an objective way of characterizing threats to the species, and thus the degree and even the kind of protection required in its management.

This entire argument has important implications for the practising restoration ecologist. The restorationist who sees himself or herself as the practical arm of some abstract, academic discipline is not likely to give great thought to the more theoretical aspects of the restoration process. Indeed, he will probably view his failures as a result of personal ignorance of the received wisdom, and as mistakes either to be forgotten or to inspire a more careful reading of the scientific literature. On the other hand, if the practising restorationist sees himself at the cutting edge of the discipline and his results as the raw data to be fed to the mill of theory, he will probably be eager to take note of, and also to contribute to, the body of information mentioned in this chapter and elsewhere in this book.

### References

Allee, W. C., Emerson, A. E., Park, O., Park, T. & Schmidt, K. P. (1949). *Principles of Animal Ecology*. Philadelphia: Saunders

Gilpin, M. E. & Soulé, M. E. (1986). The process of species extinction. In M. E. Soulé (ed) *Conservation Biology*. Sunderland, Massachussetts: Sinauer Associates.

MacArthur, R. H. & Wilson, E. O. (1967). *The Theory of Island Biogeography*. Princeton, New Jersey: Princeton University Press.

*John Cairns Jr*

University Center for Environmental Studies and Department of Biology, Virginia
Polytechnic Institute and State University, Virginia

21 | Disturbed ecosystems as opportunities for research in restoration ecology

The theme of this book is that aspects of the structure, function and dynamics of both natural and artificial systems are frequently revealed most dramatically and unmistakably when the systems are under severe stress, and during their rehabilitation and construction. This suggests that disturbed ecosystems might well be regarded not only as a major environmental problem, but also as an important opportunity for ecologists. In fact, ecologists have been concerned about the degradation or outright destruction of natural ecosystems, especially during the last 30 years. Understandably, they have spent substantial amounts of time attempting to educate society about the consequences of these assaults on the environment and to find ways to reduce, if not eliminate, them. What is curious, however, is that ecologists have taken so little advantage of the splendid research opportunities available as a consequence of sometimes legal, but almost always ethically irresponsible, destruction of the environment or catastrophic spills of hazardous materials. This is especially true in light of the value of restoration as a technique for basic research, illustrated in the preceding chapters. There is no question that disturbance theory in ecology is of interest to ecologists: a symposium with that title was sponsored by the Ecological Society of America and held at its annual meeting in Fort Collins,

Colorado, in August of 1984. The symposium was convened and chaired by Edward J. Rykel Jr, of Texas A & M University. It was well attended; presentations ranged from pure theory to application, and the question period revealed a general interest in a variety of aspects of disturbance theory.

Unfortunately, however, most biological information associated with human developments of various kinds, especially industrial development, has been obtained in the preparation of environmental impact statements (EIS) mandated by law. Such information generally consists of species inventories and various other observational data that are frequently ill prepared and almost invariably include little or no information of general interest. An uncharitable person might conclude that, despite the original intention of the legislation requiring EISs, preparation of the reports themselves has rapidly deteriorated into a ritualized fulfillment of the legal requirements. Few people read an EIS unless they are forced to do so. If comparable energy and funding were being put into studies of the structure and function of ecosystems, both under stress and following the removal of stress and during restoration, restoration ecology would be much more advanced than it now is.

The earlier chapters in this book clearly illustrate the fundamental value to ecology of well-conceived attempts to restore disturbed ecosystems – or even to construct entirely artificial systems in the laboratory for purely experimental purposes. What I would like to do here is explore the variety of disturbed systems available for study and draw attention to the kinds of research that might be carried out on them. I will also comment on some practical considerations in the partnership between scientists and industry. The question for ecologists is not whether disturbed ecosystems provide valuable information (since there is general agreement that they do). The question is why they have received so little serious attention. This question is especially puzzling – and urgent, even from a purely environmental point of view – because disturbed ecosystems are now far more abundant than undisturbed ones, and also because if some are not given attention they may adversely affect the few remaining undisturbed ecosystems.

In speculating about this imbalance, I have come to suspect that it reflects fundamental differences in the requirements of the two types of studies, that have little to do with science or the theoretical value of the information generated. The rapid and intense response of ecologists to the Mount St Helens eruption of 1980 and some other similar events elsewhere (for example, studies of Surtsee as reported by Maguire, 1977) suggests considerable interest in a significant portion of the academic community in issues pertaining to disturbance theory, recolonization and the like that can only be studied in profoundly disturbed systems. The Mount St Helens studies have already been of immense value since they showed that recolonization was much more rapid than most ecologists had predicted, raising questions about the assumptions on which the predictions were based. Nevertheless, studies of restoration ecology are likely to be carried out on sites that are not esthetically pleasing, that require interactions with other disciplines, that require ecologists to make predictions without an adequate data base, that often occur in situations where public opinion is polarized, and at worst, that may result in appearances of scientists in courts of law. Finally, one's colleagues are less likely to be impressed by the study of a garbage dump, even in an exotic location such as McMurdo Sound, than by a study of an undisturbed habitat on one of the lesser known Galapagos islands. Clearly, much valuable ecological information has been generated as a result of studies in more or less undisturbed areas. However, by now it is abundantly clear that scientifically useful information can often be gathered in communities and ecosystems that have been disturbed and are undergoing recovery or active restoration. This is true, moreover, whether the disturbance has been caused by human activities or by a natural event such as a volcanic eruption.

### Working with corporate sponsors

Since this is the case, and since it is also true that more and more scientists are chasing fewer dollars from the traditional sources, there is a real incentive for closer cooperation with funding sources that may have been less attractive when funding was plentiful. At the same time, it is important for ecologists considering this possibility to keep in mind that conditions for research in this area might be quite different from those in academia, and it would be a serious error for an ecologist, having obtained funding from, for example, a corporation sponsor, merely to proceed with research in the usual way. For a scientist carrying out research directly related to an immediate, "practical" problem, for example, it is important not only to have a falsifiable hypothesis, but also to consider the quality and quantity of information needed for a practical management decision. Whereas in academia information may be regarded as having value in itself, corporate decision makers are likely to feel that information has value only in so far as it will change a course of action. (More information about decision analysis can be found in Raiffa (1968) and Behn & Vaupel (1982).)

It is also worth remembering that deadlines are far more important in this segment of society than in academia. In the latter, colleagues who have faced the same problems are inclined to be charitable, knowing they may be the next to need sympathy. In the commercial area, postponing a decision can be exceedingly expensive and may be ruinous. As a consequence, one must be both candid and realistic when describing the types and amounts of information that will be available at particular times. This does not mean that long-term studies are not possible, but rather that the decision makers must be fully informed about what is immediately available and what is not. Finally, a researcher working on a project closely related to a practical problem may be called upon for an opinion based on what a scientist might feel is inadequate information. Decision makers in business and industry often make decisions under these conditions and generally place considerable value on expert opinion in doing so. But this is a situation in which some scientists may not feel comfortable. Ultimately, each researcher must decide the conditions under which he or she will carry out research in restoration ecology funded by organizations other than the traditional foundations and granting agencies.

A few illustrations from my own experience may help illustrate some characteristic problems. I have been called in by a corporation for the purpose of telling the management what a contractor's research means and how it applies to the needs of the organization. When this happens, it is clear that there is a serious communication problem, and it is even possible that the research bears no relationship whatever to the needs of the corporation. For example, if a corporation is required to get information about the environmental impact of an operation, the plant manager may instruct the engineer in charge of the project to employ a "local ecologist" to carry out the necessary investigation, on the assumption that involvement with a local academic institution will be good public relations. The engineer then goes to a nearby university or college and asks for an ecologist. Sometimes the engineer does not know what information is needed, and the ecologist assumes that he or she has been approached because the engineer is aware of the investigations for which the ecologist is known professionally. In fact, the engineer has no idea what an ecologist does or even whether this particular ecologist is a terrestrial or aquatic ecologist. As a consequence, the ecologist goes to the site and makes the measurements on which his or her reputation is built. The reader can write the rest of the scenario. Of course the entire fiasco could have been avoided if the engineer and the ecologist had known roughly what to ask of each other, but often this fundamental exercise in communication is either trivialized or neglected.

Fortunately, this is happening less frequently these days as each profession becomes more aware of the needs of the other.

The largest collaborative project for which I have been principal investigator involved 56 investigators in 14 departments in three colleges. Astonishingly, this particular investigation worked quite well. I will never be certain why, but I believe that two factors were of utmost importance: (1) everyone on the team was determined that the project would succeed because many of their colleagues had told them it would not; and (2) the team began interacting by discussing the size of the sampling quadrats. Surprisingly, this led to the best communication among the component study groups I have ever seen. It happened purely by accident when, having chosen an analytical technique developed elsewhere, we invited the developers to the campus to explain to the entire group how it worked.

Another persistent problem related to communication is meeting deadlines. From personal experience, I suspect that most team failures occur because an individual or a group of individuals fails to take the deadlines for completing different data-gathering activities seriously. If the study team is truly an interdisciplinary group (as opposed to a multidisciplinary group, in which the participants gather data more or less independently), the results obtained by each participant or sub-group influence what is done by the others. As a result, failure to furnish data on schedule means that the entire work plan is disrupted, and researchers who planned to work on specific dates cannot do so because the necessary information is not available. Of course there are always delays due to weather, equipment failure and the like, and schedules are always disrupted. This is to be expected. What causes members of a team to lose confidence in each other is failure to communicate the fact that problems have arisen so that the team can collectively redesign the study. Failure to acquire data merely because something else was given a higher priority can be fatal to the cohesiveness and mutual trust of a team.

The following is a list of conditions I consider most important when working as a group in collaboration with outside organizations.

☐ First, it is essential to agree beforehand that all information generated will be the property of the principal investigator and the co-investigators, to be used at their discretion in manuscripts submitted to scholarly, peer-reviewed journals. It is customary to submit information to the sponsor before publishing it. I usually encourage and receive comments from the sponsor and have frequently found these valuable. I usually send a copy of each manuscript submitted for publication to the sponsor, but do not promise to alter the contents to meet the sponsor's wishes. However, the most desirable sponsors for such research are eager for scientific credibility, and what better evidence for this than publication in peer-reviewed journals? One of the "prices" the sponsor pays for this credibility is publication of information at the discretion of the investigator, regardless of possible adverse publicity. Naturally, this credibility is even greater if the sponsor agrees to these conditions as a matter of record (i.e., in the contract or grant), before the gathering of data begins and before the nature of the evidence has become clear.

My own experience with managing the publication of results in this manner has generally been quite encouraging. For example, I once carried out a study of the ecological consequences of the accidental spilling of a fly ash pond into a river. I sent a report to the sponsor indicating that nearly all the aquatic organisms had been killed, and shortly followed this with a manuscript to a professional journal stating the same results. I also reported in the professional literature and to the citizens of the area where the spill had occurred that the recovery of the system had been astonishingly rapid, much more rapid, in fact, than anyone would have reasonably expected. This was credible because I had evidence for it and also because I had reported the unfavorable things as well. Of course, not all sponsors will act so responsibly. Elsewhere (Cairns, 1985a; in press), I have discussed some interactions that are a source of concern in this regard.

☐ With some exceptions, research carried out in collaboration with a corporate sponsor should be suitable for MS theses or PhD dissertations. The design of these dissertations is usually the result of interactions between the graduate student and the dissertation committee. The majority of the committee should not be associated with the project, though it is important for members to understand that the sponsor is supporting the research and to be aware of the obligations this entails. The challenge is to design the project so that it has scientific value and also satisfies the sponsor's requirements, and again, my experience in this regard has been quite favorable. Almost without exception, research carried out by my students under these conditions has led to publication in peer-reviewed journals.

☐ Except in exceedingly rare instances, researchers will be well advised to avoid situations where emotions are high. Occasionally, getting involved in such a situation might be justified as a public service; but in such cases, the primary goal will usually be something other than research. Generally, it will be best to keep students out of such projects and to become involved in them only as a "friend of the court". One should not support any particular point of view (i.e., side with either of the litigants) but rather be determined to get the best evidence possible so that an informed decision is possible.

☐ It is exceedingly helpful to have a liaison person representing the sponsor present throughout the study. This improves communication and makes it easier to get information about the operation, pre-site conditions, etc.

Adhering to these conditions will markedly reduce the amount of funding available – probably by 75 per cent or more, although this is difficult to document. On the other hand, ignoring them is likely to lead to a great deal of trouble and frustration – and little increase in worthwhile research.

### Disturbed ecosystems and their restoration as opportunities for basic research

Since it is virtually impossible to leave an academic institution anywhere in the world and travel very far without encountering a disturbed ecosystem, opportunities for the development of disturbance theory and restoration ecology abound, and there is no doubt that research related to such systems and their restoration can easily yield results of both practical and theoretical value. For example, a recent investigation following an industrial waste spill on the Flint River in Georgia resulted in a publication on functional groupings of protozoans (Pratt & Cairns, 1985). The Flint River research was not the only source of evidence in this study, but the information it provided was crucial. One of the advantages of this study was the presence of an impoundment (Lake Blackshear) on the river. This provided two different habitats to study. Additional evidence from the same study further elucidated the response of microbial communities to stress and is presently in review (Pratt *et al.*, unpublished data).

The basic value of well-planned studies of this kind is no accident, but arises from the very nature of the problem. Research on the recovery or restoration of disturbed ecosystems to some stable condition assumes, for example, that we have a fairly comprehensive knowledge of the structure and function of the reference (natural) systems; the recovery or restoration becomes a test of the validity of that knowledge, and also raises new questions. One benefit of studying disturbed ecosystems, for example, is that it makes it necessary to determine the characteristic variability of the model community in more detail than might otherwise be necessary. The recovery or restoration process itself provides opportunities for studying the factors influencing the stability of the system. The recolonization process is also instructive, especially when some manipulations are carried out to determine what enhances or retards recolonization.

A common charge is that "applied" projects are unsuited for long-range ecological studies, but it is worth noting that Ruth Patrick has carried out applied studies on a number of rivers for periods of time often in excess of 10 or 20 years, which have provided useful information about the species pool in such systems and the successional processes due to changes in water quality, season, etc. (Patrick, Cairns & Roback, 1967). In one such study, Patrick (1977) found evidence on the Sabine River in Texas of long-term deterioration of the diatom community. The practical benefits of these studies were early warnings of thermal or toxic stress and the detection of extremely minute quantities of isotopes of chromium that eluded conventional means of detection. From a more "basic" point of view, the benefit was an understanding of heavy metal uptake by diatoms. Similarly, the long-term studies of Robert J. Livingston of Florida State University on the Apalachicola River and its estuary have revealed some fascinating long-term cyclic effects of flooding, and have provided some extremely interesting ecological information on species interrelationships. Livingston had a wide variety of funding sources that were all skillfully fitted into a master study plan. In contrast, each of Patrick's studies was funded primarily by a single source for the entire period of study. I know of very few studies supported by traditional funding sources in which the work of a single investigator working on a single system has been supported continuously for a comparable length of time.

### Basic information from restoration project

Such work has both basic and practical value. An example from my own experience is a project I carried out with Dr James R. Pratt on a surface mined ecosystem in central Florida. Although the system being mined was originally almost entirely terrestrial, reclamation laws allowed rehabilitating the system to lakes and uplands, mostly because sufficient material was not left after mining to backfill the mine cuts. The sponsor and others contemplating similar projects received the following benefits from this study:

☐ A determination of the time required to reach a dynamic equilibrium.

☐ An assessment of the differences in the species composition of communities in reclaimed and natural lakes.

☐ Information on the range of variability of measured community responses (such as community species richness and productivity) to seasonal changes in nutrient levels, temperature, and other factors. (There were two major seasonal maxima and minima.)

☐ An evaluation of physico-chemical differences in several types of artificial lakes.

☐ Indentification of key species and species groups indicative of particular conditions.

☐ Comparison of the ecological consequence of the several reclamation methods used at the site.

At the same time, we had opportunities to test the following fundamental hypotheses in the course of the study:

☐ The structure of microbial communities depends on physico-chemical factors, and can be used to identify these factors and to evaluate their ecological significance.

☐ The nature and rate of the colonization of artificial substrates by microbial species can be used to compare conditions in different ecosystems.

☐ Although the species may vary from ecosystem to ecosystem, and although stochastic factors may play a major role in colonization, the *pattern* of colonization by species representing various functional groups tends to be similar in many ecosystems.

☐ Microbial species (especially protozoa) have broad tolerances to environmental conditions.

☐ Protozoa are widely distributed: the same species may occur anywhere that ecological conditions are appropriate.

### Some fundamental issues

The above should give some idea of the fundamental value of properly designed studies undertaken in connection with the restoration of particular disturbed sites. Now I would like to look at the matter from another direction and to consider some of the basic ecological issues that may be explored by way of restoration, provided suitable sites and circumstances can be found.

### Species and communities versus ecosystems

Many rare or endangered species are threatened with extinction because their range is naturally limited, or because their habitat has been reduced or degraded in some way. This raises the possibility that one way of increasing populations of these species is to restore disturbed areas specifically to provide habitat for them. Clearly, however, this idea immediately offers a series of fundamental challenges to ecological theory. Since the protection of ecosystems and the protection of individual species require different strategies (Cairns, 1985b), it would be an interesting exercise in both theoretical and applied ecology to find ways of restoring both functioning ecosystems and particular populations and communities simultaneously. Since there may be species that cannot be kept alive in the laboratory or in a zoo or botanical garden, but that might survive in a more complex environment under semi-controlled conditions, this is a matter of considerable environmental importance. It will certainly be a challenge to ecologists to determine whether they know enough about ecosystems to design them to accommodate particular species (see Ch. 19).

While I have suggested this as an objective for restoration on a number of occasions, I am not aware that it has ever been attempted in a serious way. Reasons for this probably include fear of not meeting legal requirements, fear of failure and fear of acquiring a permanent responsibility that could be prohibitively expensive. Of course these problems raise legal and administrative questions that are beyond the scope of this discussion: all that can be done here is to allude to the opportunities and ignore the details.

### The ecotype question

One of the unresolved legal questions pertinent to ecological restoration is whether the requirement to replace a species eliminated as a result of mining or some other disturbance is satisfied by replacing it with a population representing a different strain or race of the same taxonomic species.

This clearly raises numerous questions about the ecological importance of ecotypic variation that we are not yet in a position to answer. How important is it, for example, that ecotypes used in restoring communities closely match those present in the original or model community? From a practical standpoint a case could be made for obtaining all replacement stock, not from natural but from restored communities, since these have proved successful on the disturbed site. On the other hand, to the extent minor genetic variations are ecologically important, it might be as well to begin with closely matched stock, or to bring in appropriate stock at some stage in the restoration process so that the gene pool of the restored population closely be tolerated in the attempt to achieve various degrees of authenticity in the restoration? How closely, for example, does replacement stock have to match the original in order for structural features of the restored system (its trophic balance, for example) and functional features, such as nutrient spiralling and energy transfer, to resemble closely those of the model system? Also if these properties of the ecosystem are reproduced reasonably well, how much does it matter that the genetic composition of the component populations is rigorously authentic?

### Species substitutions

A similar question arises in considering the ecological properties and value of artificial communities quite different in species composition from those that occur naturally. The creation of such communities (which is quite common in forestry, for example, or some aspects of mine reclamation) clearly offers splendid opportunities to determine the ways species can be replaced by others that have similar functional roles. If only really authentic communities function properly or provide important ecosystem services, the ecological restraints on the restoration process will obviously be quite severe. This being the case, experiments to answer these questions should be one of restoration ecology's highest priorities.

### Time and space

It is exceedingly difficult to disentangle the ways in which temporal and spatial factors influence various events in an ecosystem. In the colonization process, for example, two identical ecosystems at different distances from the source pool of species may ultimately have the same species richness, but species accrual is generally more rapid on the system closest to the source pool. Here is a real opportunity for restoration ecology, since the rehabilitation of disturbed ecosystems can be designed specifically to untangle spatial and temporal processes. Again, it will require skill to meet both practical and theoretical needs in undertaking such studies. But it is difficult to imagine a situation in which comparable large-scale, semi-controlled experiments could be carried out in natural systems. It is also unlikely that funding for such projects would be available from the usual "ivory tower" sources.

With this in mind, it may be worthwhile to consider very briefly the range of opportunities for actually carrying out restoration ecology. Perhaps the most reliable source of support for ecosystem restoration projects is the surface mining industry. In most cases, state and federal laws require at least some degree of restoration of mined lands, which typically include both aquatic and terrestrial habitats. Moreover, since the process of mining is more or less continuous, it provides an unequalled supply of ecosystems of known age since disturbance.

Regulations requiring reclamation vary with location and jurisdiction. Coal mined lands, for example, must be reclaimed under relatively strict rules promulgated under the Surface Mining Control and Reclamation Act. This regulation applies all over the country, although most activity is in the Illinois–Pennsylvania coal belt. Laws directing the reclamation of phosphate surface mines in central Florida are less restrictive, and the development of artificial aquatic ecosystems is common in this area, amounting to more than 2000 ha annually. Since Florida regulations also allow leeway for experimental restoration, this region offers rich opportunities for manipulative as well as for descriptive and comparative research. It is worth noting that different areas offer a variety of potential sources of funding. Some states, for example, have research institutes or other state agencies that support restoration or reclamation research. Industries have an interest in funding such research since they are now paying large sums in reclamation expenses, and are eager to support work that might result in the refinement of techniques and savings in reclamation costs.

### Interactions between systems

It is taken for granted in ecology that not even the largest ecosystems are self-sufficient, and that events in one affect events in others. The well-known Hubbard Brook project furnishes one of the most striking examples of this. It is also well established that some disturbed ecosystems, such as surface mined lands, export materials that may be vastly different in both quantity and quality from those originally exported. An extreme example of this is the export from mined sites of acid water containing a variety of heavy metals that has caused severe degradation of many aquatic ecosystems. Work with disturbed systems clearly provides opportunities for studying the effects of these as well as more subtle modes of interaction between ecosystems.

## Characterizing vital signs

One way to determine when an ecosystem has been rehabilitated might well be to examine the quality and quantity of the materials exported from it. When these match the original condition reasonably closely, one could consider the system rehabilitated. Although much is known (from the Hubbard Brook, Coweeta and other studies of that type) about export processes in ecosystems, clearly much remains to be revealed. It is also worth noting that researchers at Hubbard Brook have employed both description and disturbance in their work, but that they have not yet tried experimental restoration. Since the predisturbance condition of the Hubbard Brook site is well known, it would be interesting to see if it can be restored to that condition. If theoretical ecologists do not have enough information to do this, perhaps the major research thrust of ecology should be redirected. Perhaps ecologists have been too species- and community-oriented, and have paid too little attention to ecosystems and their functioning. In any case, if the science can't predict the outcome of a restoration effort with some confidence, it clearly has a long way to go.

In conclusion, it is clear that the opportunities for basic research provided by disturbed ecosystems are very great and that they are currently under-exploited. Taking advantage of these opportunities without detracting from equally important research on pristine systems will be a challenge. It will also require ecologists to interact with government, industry and the legal system as they never have before.

### Acknowledgements

I am deeply indebted to James R. Pratt and Darla Donald for suggestions on the preparation of this chapter. The first draft of the manuscript was typed by Becky Allen and the final draft by Angie Miller. Darla Donald, Editorial Assistant, at the University Center for Environmental Studies at Virginia Tech., was responsible for overseeing the preparation of the manuscript in the form requested by the editor. The final draft of this manuscript was prepared at the Rocky Mountain Biological Laboratory in Crested Butte, Colorado, during the summer of 1985.

### References

Behn, R. D. & Vaupel, J. W. (1982). *Quick Analysis for Busy Decision Makers*. New York: Basic Books.

Cairns, J. Jr (1985a). Keynote address: Facing some awkward questions concerning rehabilitation management practices on mined land. In *Wetlands and Water Management on Mined Lands*, ed. R. P. Brooks & D. E. Samuel, pp. 9–17. University Park, Pennsylvania: Pennsylvania State University.

Cairns, J. Jr (1985b). Evaluating the options for water quality management. *Water Resources Bulletin*, **21**, 1–6.

Cairns, J. Jr (1987). Keynote address: Politics, economics, science – going beyond disciplinary boundaries to protect aquatic ecosystems. In *Persistent Toxic Substances and the Health of Aquatic Communities*. Windsor, Ontario: International Joint Commission (in press).

Maguire, B. Jr (1977). Community structure of protozoans and algae with particular emphasis on recently colonized bodies of water. In *Aquatic Microbial Communities*, ed. John Cairns Jr, pp. 355–97. New York: Garland Publishing.

Patrick, R. (1977). Diatom communities. In *Aquatic Microbial Communities*, ed. J. Cairns Jr, pp. 139–59. New York: Garland Publishing.

Patrick, R., Cairns, J. Jr & Roback, S. S. (1967). An ecosystematic study of the flora and fauna of the Savannah River. *Proceedings of the Academy of Natural Sciences of Philadelphia*, **118**, 109–407.

Pratt J. R. & Cairns, J. Jr (1985). Functional groups in the Protozoa: Roles in differing ecosystems. *Journal of Protozoology*, **32**, 409–11.

Raiffa, H. (1968). *Decision Analysis: Introductory Lectures on Choices under Uncertainty*. Reading, Pennsylvania: Addison–Wesley.

*Patricia Werner*
CSIRO Darwin Laboratories, Australia

# 22 Reflections on "mechanistic" experiments in ecological restoration

One of the basic questions of restoration ecology is: what do we really need to know about an ecological community in order to restore it? Tony Bradshaw's suggestion that restoration is the acid test of ecological understanding implies that we may have to know everything – or rather that what we need to know in order to restore a community defines what is really worth knowing about it. At the same time, of course, Bradshaw has qualified his formulation by noting that communities and ecosystems, like individual organisms, have considerable powers of self repair, so that it may well be possible to restore a community without really understanding some critical things about it. There seems to be general agreement, however, that putting communities back together again is a valuable way of coming to understand them. Furthermore, I think there is a general feeling that the ability to restore communities under various conditions and with confidence of success will turn out to be a test of ecology as a mature science.

To some extent, ecologists and restorationists are really trying to do the same thing, and it makes sense to look at what is going on in the area of restoration research from the point of view of an ecologist, and to ask, from a purely ecological point of view, just what kind of challenge restoration is. What kind of information is going to be needed actually to restore communities and ecosystems with confidence? What sorts of experiments are needed to provide that information?

Of course, one has to step back and ask what exactly are the goals of any restoration? Putting aside here such fundamental questions as what constitutes "natural", or what to do about key extinct species, or what to do if original communities were recent glacial relicts, let us focus for a moment on the ecological hierarchy. For example, an ecosystem-oriented ecologist might be concerned mainly with restoration of a certain level of productivity and/or species diversity, whereas a community ecologist might be more concerned with the "correct" relative abundances of all the various component species; and a population biologist might be more interested in restoring an endangered species, perhaps by habitat preservation. To some extent the degree of degradation sets practical limits on the viewpoint and goals of restoration. For example, the most severe reconstruction task is on areas where basic soil layers and vegetation have been removed, such as in strip mining. Here "ecosystem reconstruction" may be the only practical objective, with attention being paid to total ecosystem function rather than to particular species interactions and community structure. Less degradation might require only "vegetation reconstruction", perhaps of particular growth forms. Sites that are even less severely disturbed might provide opportunities for "community restoration", or even simply "species restoration", or reintroduction.

My own interests are in plant community development, which often takes place in areas undergoing natural succession, or active restoration. Ecologists working in this area often do experiments and are attempting to make this branch of ecology predictive. As others in this volume have pointed out, however, restoration offers a test of prediction, and so experiments in this area lead directly back to the questions raised above about what we really need to know in order to restore ecological systems, and what kinds of experiments will be needed to obtain that information.

In this chapter, I distinguish two types of experiment that may be – and in fact are being – carried out in the area of restoration, and then argue for the value of the second, more fundamentally oriented type of experiment.

It seems to me that if you consider research in restoration, the experiments generally fall into one of two groups. The first type essentially involves carrying out some kind of manipulation, making some kind of change in the environment, then stepping back to see what happens. Typically, the alterations that are made are rather gross ones that influence the entire community or ecosystem in some way – running a fire through it, for example, or adding fertilizer. Experiments of this kind are commonplace in ecological restoration and management research because they are relatively easy to carry out, and because they do provide rules of thumb for restoring and managing communities. It is important to recognize, however, that this approach essentially black-boxes many things that are going on in the community, treating them as processes for which we understand input and output without understanding the process itself in any detail. For example, it black-boxes all the species interactions, and the specific ways individual species respond to changes in the environment. This may be of great interest both to the ecologist and to anyone who is trying to restore a community.

The basic weakness of experiments of this kind is that they are not mechanistic. They may provide some information about what works under a given set of conditions, but they tell us very little or nothing about why it works. This means that you can't extrapolate from these experiments to different situations. You don't know, for instance, whether a completely different result would be obtained by running an experiment again in May instead of June. You don't know that the outcome will change completely if a key species is missing, or an extra one happens to be present. This may not matter so long as you can duplicate conditions the next time you try to get the same result – but of course this very rarely happens in ecology, since in the real world conditions vary in subtle but important ways, not just from site to site but from year to year.

The other type of experiment, of course, is the experiment designed not simply to demonstrate that a particular manipulation produces a particular result under a given set of conditions, but to provide information about why it gives those results. This type of experiment usually deals with some specific aspect of a single species or small group of species and the role these play in the community. This may involve some ecosystem function such as nutrient cycling, or it may involve the interaction between several species. It may be carried out on organisms growing alone or in small groups, or even in place in the community; but in either case, the objective is to identify and characterize mechanisms, and to get information that will help the ecologist understand why certain things happen in the system. The disadvantages of this kind of experiment are roughly comparable to the disadvantages of looking at something through a microscope: you see a great deal of detail, but your field of vision is extremely limited. Furthermore, the whole process takes a great deal of time, and there is always the chance that you are looking at the wrong thing.

This being the case, it seems to me that what we need really is both kinds of experiments. We need the rougher, more empirical experiments to help us answer urgent questions as quickly as possible for purely practical reasons;

but we also need the more mechanistically oriented experiments to provide information that will ultimately allow us to work with confidence under a variety of conditions, and even to make appropriate adjustments in those conditions to get the results we are looking for. No doubt this reflects my own bias as a scientist, but it does seem to me that it is ultimately the second, more mechanistic kind of experiment that will ultimately prove to be of the most value, not only to ecologists interested in explaining why things happen in communities and ecosystems, but also to restorationists who are trying to put those systems back together under real-world conditions. In other words, what we have to do, ultimately, is to characterize both the environment *and* the responses and utilization curves of particular organisms or functional groups of organisms. *Then*, with both of those things done, whether we are trying to recolonize a sand dune, or to put species back into a particular forest, we will be able to predict the behavior and the interactions of the species or species groups, and on the basis of this understanding, also to predict the outcomes of those interactions.

I would like to illustrate this point with two examples drawn from my own research. Though neither of these projects was carried out with restoration of a community in mind, they both involved the experimental construction, or partial construction of communities in order to test an hypothesis about community dynamics. Both led to information that I would hope might be of value to someone who is actually trying to restore a community on a disturbed site in the field.

The first project was an attempt to find out something about how competition influences the composition of plant communities (Werner, 1979; Werner & Platt, 1976). The species I selected for this work were two species of goldenrod: early western goldenrod (*Solidago missouriensis*) and Canada goldenrod (*S. canadensis*). Both of these species grow in prairies, but under quite distinct conditions, early goldenrod occurring almost exclusively on the dry ridges, and Canada goldenrod almost exclusively in the wetter depressions. The classic descriptive studies of prairies told us this much. What they did not tell us was why these species distribute themselves in the community in this way. To find out, we carried out a series of experiments to determine the physiological tolerances of each species at various points across the topological or edaphic gradient, as well as each species' response to competition with other species in the community. What I found was that early goldenrod survived across the entire gradient as long as competing plants were removed – in fact, the wetter the soil the larger the individuals grew. But when the rest of the plant community was left in place, and individuals of early goldenrod were added in either seed form or as small transplants, they were not able to survive at the wetter, more productive end of the gradient. So, although this goldenrod prefers the wetter sites physiologically, in the wild it is restricted to the drier sites by competition. The story is quite different for the Canada goldenrod, however. This species did not survive on dry sites, regardless of whether or not there were competing plants, and on the wet sites individuals grew equally tall with or without surrounding neighbors. Hence, in the wild, Canada goldenrod is apparently found only at the wetter end of edaphic gradients both because of its physiological intolerance to conditions on ridges, and also because of its ability to compete effectively with vegetation in the depressions. In separate greenhouse experiments these two goldenrod species were joined by two others – gray goldenrod (*S. nemoralis*) and grass-leaved goldenrod (*S. graminifolia*) – and grown alone or in various combinations across artificially constructed gradients of water-holding capacity. This turned out to be most instructive, since the results showed a continuum among these four species, with the growth of all increasing with soil wetness as long as other species were not present. Each species, however, had a different tolerance range in relation to the drier end of the gradient, with gray goldenrod growing well even in the very driest area and early, Canada, and finally grass-leaved goldenrods showing, in that order, decreasing tolerance to drier conditions. Furthermore, when all four species were grown together, there was a distinct sorting into zones, with the grass-leaved goldenrod taking its physiological, wet-end zone and the other species in turn being displaced by competition from their wetter-end physiological ranges and taking their drier-end range where they could survive and their wetter-end neighbors could not.

The important point is that the microzones found in the prairie are not a result of physiological tolerances to abiotic factors alone; indeed, all species preferred the intrinsically more productive end; but the species had various competitive abilities that seemed to be traded off somehow with tolerance to poor conditions. What we discovered next was that the critical factor that accounts for the distribution of these species within the community is the way individual plants allocate resources early in life. The dry-tolerant, low-production species (e.g., gray goldenrod) allocate relatively little to new leaves and much to root tissue in contrast to the dry-intolerant, high production species (e.g., grass-leaved goldenrod) which allocate a high proportion of new photosynthate to new leaves and a taller stem, and relatively little to root tissue. Physiological studies have shown no differences in photosynthesis rates or water use efficiencies on a per leaf-area basis between these species. On a whole-plant basis, however, the latter species requires more water, and in turn has a higher overall growth rate on moist sites and can therefore outcompete slower-growing, water-conserving species under these conditions.

Having identified this factor, we are now in a position to provide a small, but I hope useful, piece of advice to the prairie restorationist who wants to be sure that these goldenrod species are present in a restored prairie. In the beginning of a restoration, before vegetation has developed very far, many species will probably appear more successful than they will ultimately be. That is, individual plants on rich and/or wet areas may grow better than other individuals of the same species growing on poorer sites. Given time, however, as neighbors grow and competitive sorting takes place, species will probably be displaced toward the poorest portions of their physiological range. This is because each species will be limited at the preferred, richer or wetter end of its range by competition; but it will be limited at the drier or poorer end of its range only by its physiological tolerance to the poorer growing conditions there. Hence, a restorationist must be careful to plant each species on sites at the *upper* range of its tolerance. Even if initially it seems as though growth is poor there, this is likely to be the only place the species will survive in the fully restored community.

Information of this kind may prove to be quite useful to the practising restorationist. What is of special interest in the context of this book, however, is not only that these experiments produced results that may be useful, but that they are examples of restoration ecology. That is, even though they were not carried out in reference to the actual restoration of a native community, they did involve reconstruction of a community – not whole-sale, but in pieces. Indeed, this may be what one would expect of restoration or "synthetic" experiments carried out to test specific, fundamental hypotheses about community structure and function. It may be that experiments of this sort suggest the link that might be developed between field and laboratory research in this area. Work of a rather similar nature has been described by Kay Gross in Chapter 12.

My second example is a field experiment designed to reveal something about species combinability, and the factors that determine the ability of plant species to coexist (Werner, 1977). From the outset we hoped that these would include broad-scale, easily identifiable characteristics, such as plant life form or architecture. As in the experiments described above, our experiments involved addition of plants to communities. In this case, we sowed known numbers of teasel seed (*Dipsacus sylvestis*) in eight adjacent small fields in Michigan that were undergoing succession following agricultural use. Teasel was initially absent, but only as a result of lack of dispersal, and the eight fields chosen had similar histories and differed only in species composition. Following a period of growth, we then attempted to determine whether the success of teasel in colonizing and growing on various sites was related to the other species present. The patterns that emerged from this effort were very obscure, however. This was because each teasel plant had as neighbors approximately six plants of any of more than 100 species, and since most species in a plant community are rare, the combinations of species surrounding the teasel plants were virtually endless. When neighbors were grouped by either life form or architecture, however, the behavior of teasel made sense. Teasel, a biennial plant that grows as a rosette in its first year, grew well with grasses, and in fact added to the grass community without any decrease in productivity of either teasel or grasses. But growing with other dicotyledonous plants, teasel grew more slowly so that it remained in a rosette phase for two years. Also, total community productivity was not increased in this case, teasel replacing some of the biomass of the indigenous community. Shade from shrubs also slowed teasel growth, but over a grass understory teasel added in easily, whereas over a dicot understory it grew so slowly that it failed to reach reproductive size before dying out. The results indicated that it was not any single species, or even a few particular species, that prevented colonization of teasel, but certain broadly defined *types* of species, and that these types can conveniently be combined when one is attempting to deal with an entire plant community. Ongoing work is being carried out by Deborah Goldberg at the University of Michigan to explore the hypothesis that plants are equivalent in their competitive abilities on a per-biomass basis, and that amount of biomass (size) results in superiority. Early results indicate this is likely within certain groups of species, and that these groups conform roughly to growth forms.

All of these studies suggest that, contrary to what one might expect on the basis of studies that emphasize slight differences in the requirements of different species, the shape of plants – where they place their absorbing surfaces, both above-ground and below – may very often be a factor of overriding importance in the shaping of a plant community. Minor differences between species are important whenever plants are actually in contact with each other; but of all the possible species pairs in a field, all degrees of difference can be found; and so the total effect on a given species depends upon the relative abundances of many other species. It seems, however, that in many instances these numerous species can be considered in terms of a few simple growth forms, a useful and easily indentifiable characteristic.

This is obviously of considerable interest both from a theoretical point of view, and also from the point of view of restoring a plant community. One of the very basic questions about restoration is: just how difficult is it? Also, to what extent is it going to be necessary to duplicate species composition precisely in order to be considered successful? To what extent will it be possible to substitute species, either for convenience, or for some practical or economic reason, or simply because a species that may have been a component of the native community has become extinct or is unavailable for some other reason. Our results indicate a great deal of plasticity in this matter, suggesting that it is often quite easy to replace one species with another within a community, and that in many cases all you have to worry about is the growth habit and ultimate shape of the plant, rather than a mass of detailed information about the physiological differences among various species.

This, in turn, suggests that the complexity of the task of restoration will vary greatly depending on whether one is trying to reconstruct an ecosystem or actually to reassemble an authentic community, complete with the appropriate species. In fact, if we are not committed to the reestablishment of particular species–species interactions, and are prepared to be content with a certain level of species diversity or productivity, then the rules for reconstructing the system may actually be quite simple. It is only when one begins to be concerned with the precise reproduction of whole, meticulously authentic communities that the problem begins to look extremely complicated.

This may help to explain why those who think about the landscape in terms of ecosystems tend to be a bit more optimistic about our chances of restoring disturbed systems than those whose work is more community oriented.

### References

Werner, P. A. (1977). Colonization success of a "biennial plant" species: experimental field studies in species colonization and replacement. *Ecology*, **58**, 840–9.

Werner, P. A. (1979). Competition and co-existence of similar species. In *Topics in Plant Population Biology*, ed. O. T. Solbrige, S. Jain, G. B. Johnson & P. A. Raven, pp. 287–310. Columbia: Columbia University Press.

Werner, P. A. & Platt, W. J. (1976). Ecological relationship of co-occurring goldenrods (Solidago; Compositae). *American Naturalist*, **110**, 959–71.

*Jared Diamond*
University of California-Los Angeles

# 23 | Reflections on goals and on the relationship between theory and practice

It is a pleasure to have this opportunity to summarize what I see as the main questions and conclusions facing restoration ecologists. Before going on to comment on these matters, however, it might be well for me to describe briefly the perspective from which I view the subject of restoration and restoration ecology. My perspective is twofold: it is partly that of an ecologist who has been working for many years helping to set up national park systems for the governments of Indonesia and several other Pacific countries, and the World Wildlife Fund; it is also partly that of a physiologist working in a school of medicine, another healing profession that shares with restoration the problems of deciding on goals, calculating what we really need to know to be successful, and compromising with practical constraints.

In reflecting on the preceding chapters, it seems to me that we have been focusing on three main questions: what the explicit goals of restoration should be; what is the minimum knowledge needed to achieve those goals; and finally, accepting the premise that restoration not only takes from ecology but also gives to it, we have been considering just what community ecologists can learn as a result of attempts to restore disturbed ecosystems.

Of these three questions, let me reflect first on the last one, which stands a little aside from the other two. What is it that ecology can learn from restoration? From our deliberations, two sorts of lessons have emerged: one is that the traditional approach of ecology is reductionist. We approach the understanding of natural communities by studying certain pieces of them, or by studying certain properties of them. As several authors have pointed out, restoration provides new knowledge not offered by traditional approaches, because restoration is synthetic. It attempts to build communities, and that's justifiably considered the acid test of one's understanding, as Dr Bradshaw has suggested.

This is true because the community you are trying to restore generally won't function until you've supplied all of the necessary elements, including those that nature normally provides but that you may forget about or actually be unaware of. Again, to take an example from Dr Bradshaw, he showed us a slide of a waste heap which, it turned out in the second year of restoration, had been unsuccessfully restored. The reason was an unsuspected need for fertilizer to provide nitrogen that would normally have been provided by lichens. Thus one thing that such restoration efforts can offer to ecology is that unique insight provided by a synthetic approach, which complements the reductionist approach to ecological research.

But there's another set of lessons that restoration can offer ecologists, and that has to do with the severe limitations of the field experimental approach in ecology. There is no doubt that the controlled experiment with preselected perturbations and randomized experimental design is the most powerful tool of ecology, just as it is the most powerful tool of the other sciences. In ecology, however, most of the perturbations that would yield far-reaching insights are either immoral, illegal or impractical. Our field experiments usually run for a very short time – rarely more than a few years – on tiny spots rarely as big as a hectare. Even within these small areas we are confined to those small perturbations, those modest additions or subtractions that we can carry out without being persecuted by our governments, our neighbors, or our consciences.

Think, for example, how much faster community ecology would progress if you graduate students here today weren't restricted to the minuscule field experiments that are now in favor, but if each of you were permitted to select and burn some part of the city of Madison twice a year; or if you could reintroduce wolves into an area of your choice, exterminate the local population of a select species, or dredge and flood a Wisconsin farm and convert it into a marsh. In fact, there is only one way that you get to carry out big manipulations of this sort, and that's by getting involved in restoration projects and taking advantage of the opportunities they offer. These projects vastly expand the spatial and temporal scales of ecological work. They also allow us to work with perturbations far greater than we would be able to work with in any other way. For all these reasons they vastly expand the range of research we can do, the ideas we can test, and the conclusions we can hope to draw from our work.

Those contributions, I think, have emerged from our discussions as the two major contributions of restoration to the science of ecology. Taken together, these two contributions – the synthetic approach and the expanded range of perturbations – are so obvious and so important that I would argue that some training in restoration ecology surely should be part of the training of any ecologist in the future. In short, restoration ought to play the same role in the training of ecologists that practical work in the laboratory plays in the training of organic chemists.

The second of the questions on which our discussions have focused during the last two days is a more difficult one. That is the question, what should the explicit goals of restoration be? This is something restorationists discuss a lot. Words most often used in this volume have been the words "natural" and "self-sustaining". We've heard that the goal of restoration should be to recreate a natural community, or to recreate a self-sustaining community, or perhaps to preserve a community for posterity in essentially its natural state. This goal – this formulation of the goal — isn't a self-evident mandate, however. It is a choice based on values, and it is only one of many possible choices. We know that different people with different values would make different choices about the same site. And even if we adopt the goal of a natural community, it is important to recognize that this goal is extremely ambiguous and needs to be defined very carefully in order to be useful.

The problem is with the very idea expressed in the word "natural". What do we mean by this word? Do we mean untouched by human beings? In fact, almost every place on earth, except the polar regions and a few remote islands, had been severely influenced by *Homo sapiens*, directly or indirectly, long before the European colonization of the world took off in 1492. Which condition is the "natural" condition we are trying to create? In America, for example, is it the condition of the landscape at the moment Europeans arrived? Or the condition that applied a hundred years earlier – or five hundred? Or perhaps three or four centuries later, in 1792, or 1892? Clearly, none of these conditions has more right to be considered natural than any of the others. And we have not yet even considered the influence of American Indians. In some areas it was very great. Is this to be taken into consideration or discounted?

Our decisions in this matter will necessarily be somewhat arbitrary. Moreover, even if we could agree on a definition of "naturalness", none of us wants to restore every community to its natural state. Most of us would applaud restoring Curtis Prairie to its natural state. At the same time, most of us would agree that completely different goals are appropriate in other situations. We would agree, for example, that a farmer making a fish pond should stock it with species selected for maximum productivity rather than with species chosen because they represent the native fauna of Wisconsin as of, say, 1790. And we would agree that a highway engineer should stabilize an embankment with whatever plants do it best, native or otherwise.

In many other cases we might prefer a natural community but have to settle for one that is incomplete or only partly natural because the former natural community, whatever it was, is unattainable. We talk as if this were only an occasional problem. Actually, I think it will turn out to be the rule rather than the exception for at least four reasons. One reason has to do with the statement of Aldo Leopold: *To keep every cog and wheel is the first precaution of intelligent tinkering.* In restoration this is a noble but unattainable goal because many species have already been exterminated. We can't restore ciscos to the Great Lakes, passenger pigeons to our oak forests, moas to New Zealand, or thousands of other plants and animals native to other communities around the world. Some of those plants and animals were of major ecological importance, but they are now extinct.

A second reason why we often can't restore fully authentic natural communities is the presence of introduced species, which would be difficult or impossible to eradicate. It would be nearly impossible, for example, to exterminate starlings and European weeds from the United States, lampreys from the Great Lakes, rats from oceanic islands, deer and opossums from New Zealand, and Nile perch from Lake Malawi in Africa. All of these are introduced species that have had a tremendous effect on the communities they invaded.

Still a third reason why we can't restore natural communities is the problem of scale. We're not restoring the natural landscape but tiny pieces of it. The integrity of these tiny pieces will inevitably depend on influences from outside their borders. Hence restored communities usually cannot be self-sustaining. They'll have to be managed, we humans supplying the inputs that were formerly provided from beyond the boundaries.

For example, Kay Gross has pointed out that a population isolated in a small area may be highly unstable. Restored communities tend to be small, and to support small populations prone to extinction partly because they depend on resources beyond the boundaries of the community. We are the ones who will have to supply those resources to small, isolated communities to prevent extinctions and so to preserve the community itself. And doing this is a form of restoration – or at least active maintenance.

Ultimately, it doesn't even matter much whether the community is preserved or restored. For example, during my visits to Africa, I have come to realize that the most painful lesson that national park planners there have had to learn is that park boundaries severed the annual migration routes of some of the big mammals, which have a big impact on the parks. This is true, for example, of the elephants of Tsavo National Park. It is also true of the wildebeests of Kruger National Park, and as a result biologists had to give up their idea of considering even huge Kruger as a self-sustaining entity. They have instead had to manage Kruger once the migration routes were severed.

Take another example from our discussion here: prairie patches, whether natural or restored, depend on fires arising beyond their boundaries. But the city of Madison obstinately and disgracefully refuses to let fires in the city just follow their natural course. Hence it's the Arboretum that has to supply its own fires in order to maintain its prairies.

There is another question here, and that is the purely practical question of economics. Even assuming that we could agree on what an authentically restored community would be, and that we knew how to restore it, how much should we invest in achieving this goal? There will be many cases where this will be a legitimate subject for debate. Just what goals are appropriate, given finite budgets, limited time and perhaps inadequate restoration techniques? Any healing art faces this question. In medicine, for example, some victims of heart and kidney disease could be helped by extremely expensive treatments. But how many such patients, and which ones, should receive this expensive treatment? In the case of prairies, for example, should we use prairie restoration funds to weed out exotic plants that are structurally compatible with prairies? Should we instead leave those exotics there and use the money to go restore some other prairie? Or what about the reclamation of phosphate mines in Florida? Wayne Marion and Timothy O'Meara showed us in their poster exhibit (see *Restoration & Management Notes* 3:34) that bird diversity is lower on reclaimed land than on adjacent areas that had been similarly disturbed and then left unreclaimed. This raises a question. What is our goal on these lands? If it is to increase or maximize the diversity of bird species, we are obviously wasting our money – perhaps in an attempt to meet some unrealistic goal of restoration.

The point that I'm trying to make about the goals of restoration can be summarized in this way. If we claim, as we often have in the course of this volume, that the goal is to restore communities to their natural state, or to a self-sustaining condition, we are proposing a fictitious goal that can only create conflict among ourselves, and conflict between us on the one hand and the government and public on the other.

There are simply too many problems and ambiguities about such a goal. Over most of the surface of the earth there are no really "natural", undisturbed communities left. If we arbitrarily define natural to mean "as first seen by Europeans", we face major ambiguities. Even if we did define some nineteenth century condition as the "natural" condition, there are many communities whose species composition we couldn't restore even with an infinite amount of time and money. There are many other communities where even the most ardent environmentalist would adopt some goal other than restoration to the natural state. Finally, even for those communities where we might like to achieve a natural, self-sustaining state, practical considerations force us to lower our aspirations. Yet all these manipulations of communities – to achieve a natural state, a partly natural state, a wholly artificial state – all these manipulations belong to the same applied science. This is implicit, for example, in John Harper's chapter on agriculture and its relationship to ecology. We should not divide this spectrum of manipulations at some arbitrary point and say that the people on the two sides of this arbitrary point should form separate departments, go to separate buildings, and consider themselves as belonging to separate disciplines. Therefore, I see it as a major conceptual task of restoration to spare us the conflict created by fictitious goals and to decide what our real goals are – or, more specifically, to decide on a whole spectrum of real goals appropriate to the spectrum of situations we will have to deal with.

That, then, is the second of the three questions that have come up in the course of our discussion of ecological restoration and restoration ecology. The third of our questions, and it is a difficult one, is what's the minimum knowledge that we restorationists need in order to achieve our goals? Clearly it would be nice to know the detailed ecology of every species and its interactions in the community that we are going to restore. Now that we've heard the detailed accounts of Mike Gilpin on the ecology of *Drosophila* communities in glass bottles, or Kay Gross's account of the ecology of plants in successional fields, I would hate to have to restore any of those communities without at least as much information as Kay or Mike provided us. Restorationists really will find, I think, that it is realistic to seek such detailed ecological information in restoration projects carried out under favorable conditions, where there is plenty of money and a fair amount of time, and the goal is to restore the genetic diversity of small pieces of endangered habitat with a modest number of species. This may be the case, for example, in restoring Wisconsin prairies, or perhaps in restoring the endangered native bird communities of Hawaii.

In other cases, though, there are so many species, and we know so little about the ecology of most of them, that we will obviously have to settle for much less. New Guinea rainforests, for example, have about 600 bird species, 200 mammal species, and thousands of plant species, including unknown numbers of undescribed ones. Even for most of the described species, we know nothing about their reproductive biology. Thus it is clear that restoration will often be a crisis discipline that allows no time at all for background ecological studies.

For example, in the mid-1970s the government of Papua New Guinea, told its ecologists, "We, the government, together with a timber company, have already begun clear-cutting the rainforests of the Gogol River. Secondary growth is starting to spring up and you may begin restoring those rainforests right now".

Another example: in 1979, the Indonesian government told the World Wildlife Fund, myself and other consultants, "Next year we, the Indonesian government, are going to award the timber leases in Indonesian New Guinea. In six months give us the finalized national park plan for Indonesian New Guinea". The problem was that large areas of Indonesian New Guinea had never even been entered by biologists. We gave them advice within six months, but obviously some of that advice was based upon pure guesses, even about what species occurred where.

These demands for instant advice are not unusual. In many parts of the world they are the rule. The government will say, "We want to plan a national park in this area of 3000 square miles. A biologist has never been there. Here are two weeks and $1000. Find out what is there and what we need to plan the national park, conserve it, and restore any damaged areas". We are going to be seeing this over and over again in the future. Hence, it seems to me that the most important problem that restoration ecologists face is: given some limited amount of time and money, what is it that we most urgently need to know in order to restore – and manage – a community? Just what ecological information should we try to acquire for that money and within that time? Also what are the details that we will ignore and package into a black box for which we will settle on knowing only the input and the output?

We've seen examples of this approach here in this volume; for instance, John Aber's measurements of soil mineralization within plastic bags. John has proceeded in this way because he is convinced that for certain purposes it is important to know the rate of mineralization, and that one can get along without knowing exactly how it takes place. To me, confronted with these two-week deadlines in Indonesian New Guinea, this seems to be a good example of how to use time and money efficiently. However, these examples may cause some dissent arising from the differing needs of pure science and applied science.

Let's make clear, however, what the differing goals and methods of pure science and applied science are, and let's make clear what we as ecological restorationists are doing. In pure science the goal is to understand the world. There are no time limits. The definition of the scientific method is to make general statements that are supported by adequate evidence. Scientists are judged by the connection between their evidence and their conclusions. Thus, if your conclusions are based on inadequate evidence, you are simply practicing bad science.

In applied science, on the other hand, the goal is to solve a practical problem, using whatever scientific information is available. There are time limits. You are judged by the quality of your results, not by the quality of your reasoning and the scientific evidence that led you to those results. If your methods and reasoning don't lead you to good results, that's what makes you a bad applied scientist.

In doing restoration, we are doing applied science, not pure science. In the process we will continually be asking ourselves what is the information that we really need in order to get high-quality results, given the constraints on our information-gathering ability.

Just to summarize, then, I see four conclusions as having emerged from our discussion. First, restoration has two important things to offer to ecology. It offers the opportunity to learn by synthesizing communities rather than reducing or analyzing them; and it offers the opportunity to carry out major experimental manipulations, ones that ecologists would never be permitted to carry out under any other circumstances. At the same time, restoration ecology now faces two major conceptual challenges. One is to decide what we really mean by our goals when we pretend that we are restoring natural, self-sustaining communities – which we rarely, if ever, are. The second is to decide, given limited amounts of time and money, what we most urgently need to know in order to achieve our goals. These are important, urgent questions for restoration that clearly have important implications for the science of ecology as well.

# Index